Methods and Protocols in Food Science

Series Editor
Anderson S. Sant'Ana
University of Campinas
Campinas, Brazil

For further volumes:
http://www.springer.com/series/16556

Methods and Protocols in Food Science series is devoted to the publication of research protocols and methodologies in all fields of food science.

Volumes and chapters will be organized by field and presented in such way that the readers will be able to reproduce the experiments in a step-by-step style. Each protocol will be characterized by a brief introductory section, followed by a short aims section, in which the precise purpose of the protocol will be clarified.

Foodborne Pathogens

Loop-mediated Isothermal Amplification Protocols

Edited by

Alejandro Garrido-Maestu

Laboratory of Microbiology and Technology of Marine Products (MicroTEC), Institute of Marine Research (IIM), CSIC, Eduardo Cabello, Vigo, Spain

Alexandre Lamas

Department of Analytical Chemistry, Nutrition and Bromatology, Universidade de Santiago de Compostela, Lugo, Spain

Springer

Editors
Alejandro Garrido-Maestu
Laboratory of Microbiology and Technology of Marine Products (MicroTEC)
Institute of Marine Research (IIM)
CSIC
Eduardo Cabello, Vigo, Spain

Alexandre Lamas
Department of Analytical Chemistry, Nutrition and Bromatology
Universidade de Santiago de Compostela
Lugo, Spain

ISSN 2662-950X ISSN 2662-9518 (electronic)
ISBN 978-1-0716-4525-3 ISBN 978-1-0716-4526-0 (eBook)
https://doi.org/10.1007/978-1-0716-4526-0

This Springer imprint is published by the registered company Springer Science+Business Media, LLC, part of Springer Nature.
The registered company address is: 1 New York Plaza, New York, NY 10004, U.S.A.

Preface to the Series

The Methods and Protocols in Food Science series is devoted to the publication of research protocols and methodologies in all fields of food science. The series is unique as it includes protocols developed, validated, and used by food and related scientists, and theoretical basis is also provided for each protocol. Aspects related to improvements in the protocols and adaptations and further developments in the protocols may also be approached.

The Methods and Protocols in Food Science series aims to bring the most recent developments in research protocols in the field as well as very well-established methods. As such the series targets undergraduates, graduates, and researchers in the field of food science and correlated areas. The protocols documented in the series will be highly useful for scientific inquiries in the field of food sciences and presented in such a way that the readers will be able to reproduce the experiments in a step-by-step style.

Each protocol will be characterized by a brief introductory section, followed by a short aims section, in which the precise purpose of the protocol is clarified. Then, an in-depth list of materials and reagents required for employing the protocol is presented, followed by comprehensive and step-by-step procedures on how to perform that experiment. The next section brings the dos and don'ts when carrying out the protocol, followed by the main pitfalls faced and how to troubleshoot them. Finally, template results will be presented and their meaning/conclusions addressed.

The Methods and Protocols in Food Science series will fill an important gap, addressing a common complain of food scientists, regarding the difficulties in repeating experiments detailed in scientific papers. With this, the series has a potential to become a reference material in food science laboratories of research centers and universities throughout the world.

Anderson S. Sant'Ana

Preface

Loop-mediated isothermal amplification, known by its acronym LAMP, is nowadays the most commonly applied isothermal nucleic acid amplification technique. Since the initial publication of its mechanism back in 2000 by Notomi et al., LAMP has emerged as powerful tool in many fields. This is mainly due to several key factors. Like PCR, and real-time PCR, it can specifically amplify a specific target within a complex mixture, thus allowing to significantly reduce the turnaround time of culture-based microbiological methods. In addition to this, it presents several key features such as its higher resistance to typical *Taq* Polymerase inhibitors, compatibility with a plethora of naked-eye detection strategies simplifying result interpretation, and its isothermal nature allows to reduce the investment needed to set it up in a laboratory, contrary to other molecular biology-based methods. LAMP-based methods experienced a boost, like many other isothermal techniques, during the COVID-19 pandemic due to the possibility of being performed in decentralized setups, and/ or its potential implementation in different types of Point-Of-Care (POC) devices, given the attributes just mentioned.

In this book the reader will find up-to-date, LAMP-based methodologies for the detection and characterization of relevant, human, foodborne pathogens. As such, it is divided into five major sections. It begins with a general overview about LAMP basics. This is followed by eight chapters focused on well-known, regulated bacterial pathogens. Then, four chapters dedicated to virus detection; and another three focused on "other microbes." The last four chapters are dedicated to "novel applications" where the LAMP is applied in a nonclassical way by its combination with nanomaterials, bacteriophages, among others.

As a whole, the book provides step-by-step protocols for the implementation of this novel technique in any given laboratory, regardless of the experience of the personnel, as well as the infrastructure available. This book is an ideal text for all those interested in getting started in LAMP.

Eduardo Cabello, Vigo, Spain *Alejandro Garrido-Maestu*
Lugo, Spain *Alexandre Lamas*

Contents

Contributors

SARA ABALDE-CELA • *International Iberian Nanotechnology Laboratory (INL), Braga, Portugal*

VANTARAKIS APOSTOLOS • *Department of Public Health, Medical School, University of Patras, Patras, Greece*

CHRISTINA ANN BARNES • *Food Safety, Neogen Corporation, Lansing, MI, USA*

JULIA D. BECHTNER • *Department of Food Science, Aarhus University, Aarhus, Denmark*

RAFAEL DORIGHELLO CADAMURO • *Microbiology Division, Faculty of Sciences, University of Burgos, Burgos, Spain; Research Centre for Emerging Pathogens and Global Health, University of Burgos, Burgos, Spain; Department of Microbiology, Immunology and Parasitology, Federal University of Santa Catarina, Florianópolis, SC, Brazil*

GUNIMALA CHAKRABORTY • *Department of Molecular Genetics &Cancer, Nitte University Centre for Science Education and Research, Nitte (Deemed to be University), Mangalore, Karnataka, India*

BENJAMIN WAI KIT CHAN • *PocNAT Limited, Hong Kong, China*

CHANGSUN CHOI • *Department of Food and Nutrition, College of Biotechnology and Natural Resources, Chung-Ang University, Anseong, Republic of Korea; Department of Food and Nutrition, College of Biotechnology and Natural Resources, Chung-Ang University, Seoul, Republic of Korea*

KELLY J. DOMESLE • *Center for Veterinary Medicine, U.S. Food and Drug Administration, Laurel, MD, USA*

MARIANA ALVES ELOIS • *Microbiology Division, Faculty of Sciences, University of Burgos, Burgos, Spain; Research Centre for Emerging Pathogens and Global Health, University of Burgos, Burgos, Spain; Department of Microbiology, Immunology and Parasitology, Federal University of Santa Catarina, Florianópolis, SC, Brazil*

GISLAINE FONGARO • *Department of Microbiology, Immunology and Parasitology, Federal University of Santa Catarina, Florianópolis, SC, Brazil; Laboratory of Applied Virology, Department of Microbiology, Immunology and Parasitology, Federal University of Santa Catarina, Florianópolis, Brazil*

ALEJANDRO GARRIDO-MAESTU • *Food Quality and Safety Research Group, International Iberian Nanotechnology Laboratory, Braga, Portugal; Labotaroty of Microbiology and Technology of Marine Products (MicroTEC), Institute of Marine Research (IIM), CSIC, Vigo, Spain*

BEILEI GE • *Center for Veterinary Medicine, U.S. Food and Drug Administration, Laurel, MD, USA*

SAIOA GÓMEZ • *University Rovira i Virgili, Biochemistry and Biotechnology Department, Nutrigenomics Research Group, Tarragona, Spain; Pere Virgili Institute for Health Research (IISPV), Tarragona, Spain*

PATRICK HARDINGE • *School of Biosciences, Cardiff University, Cardiff, UK*

MARWA M. HASSAN • *University of Surrey, Royal Veterinary College London, University College London, London, UK*

SARA HEALY • *University of Surrey, Royal Veterinary College London, University College London, London, UK*

MD IQBAL HOSSAIN • *Department of Food and Nutrition, College of Biotechnology and Natural Resources, Chung-Ang University, Anseong, Republic of Korea*
LIJUN HU • *Office of Laboratory Operations and Applied Science, Human Foods Program, U.S. Food and Drug Administration, College Park, MD, USA*
SOONTAG JUNG • *Department of Food and Nutrition, College of Biotechnology and Natural Resources, Chung-Ang University, Seoul, Republic of Korea*
KEZIA KOZEL • *University of Surrey, Royal Veterinary College London, University College London, London, UK*
HYOJIN KWON • *Department of Food and Nutrition, College of Biotechnology and Natural Resources, Chung-Ang University, Anseong, Republic of Korea; Department of Food and Nutrition, College of Biotechnology and Natural Resources, Chung-Ang University, Seoul, Republic of Korea*
ALEXANDRE LAMAS • *Food Hygiene, Inspection and Control Laboratory, Department of Analytical Chemistry, Nutrition and Bromatology, University of Santiago de Compostela, Santiago, Spain; Food Hygiene, Inspection and Control Laboratory (Lhica), Department of Analytical Chemistry, Nutrition and Bromatology, Veterinary School, Campus Terra, Universidade de Santiago de Compostela, Lugo, Spain*
ERIC WAI HIM LEE • *SGS Hong Kong Limited, Hong Kong, China*
THOMAS MING HUNG LEE • *PocNAT Limited, Hong Kong, China; Department of Biomedical Engineering, The Hong Kong Polytechnic University, Hong Kong, China*
LORENA CASADO MARTIN • *Microbiology Division, Faculty of Sciences, University of Burgos, Burgos, Spain; Research Centre for Emerging Pathogens and Global Health, University of Burgos, Burgos, Spain*
RYAN C. MCDONALD • *Center for Veterinary Medicine, U.S. Food and Drug Administration, Laurel, MD, USA*
MASMARIKA MOHAN • *Department of Molecular Genetics &Cancer, Nitte University Centre for Science Education and Research, Nitte (Deemed to be University), Mangalore, Karnataka, India*
JULIET MOHANRAJ • *Department of Infectious Diseases, Nitte University Centre for Science Education and Research, Nitte (Deemed to be University), Mangalore, Karnataka, India*
LUDWIG NIESSEN • *Chair of Microbiology, TUM School of Life Sciences, Technical University of Munich, Freising, Germany*
BERNARDO A. NOGUEIRA • *International Iberian Nanotechnology Laboratory (INL), Braga, Portugal*
JORGE SANTAMARÍA PALACIOS • *Microbiology Division, Faculty of Sciences, University of Burgos, Burgos, Spain; Research Centre for Emerging Pathogens and Global Health, University of Burgos, Burgos, Spain*
JUAN LUIS PARIS • *Instituto de Investigación Biomédica de Málaga y Plataforma en Nanomedicina-IBIMA Plataforma BIONAND, Málaga, Spain*
NEIL PERCY • *Food Safety, Neogen Corporation, Lansing, MI, USA*
DANIEL PÉREZ-ALONSO • *Microbiology Division, Faculty of Sciences, University of Burgos, Burgos, Spain; Research Centre for Emerging Pathogens and Global Health, University of Burgos, Burgos, Spain*
MARTA PRADO • *Food Hygiene, Inspection and Control Laboratory, Department of Analytical Chemistry, Nutrition and Bromatology, Campus Terra, University of Santiago de Compostela, Santiago, Spain*

DAVID RODRÍGUEZ-LÁZARO • *Microbiology Division, Faculty of Sciences, University of Burgos, Burgos, Spain; Department of Microbiology, Immunology and Parasitology, Federal University of Santa Catarina, Florianópolis, SC, Brazil*

LAURA RODRIGUEZ-LORENZO • *International Iberian Nanotechnology Laboratory (INL), Braga, Portugal*

FOTEINI ROUMANI • *Laboratory of Food Chemistry, Department of Chemistry, School of Sciences, National and Kapodistrian University of Athens Zografou, Athens, Greece; Laboratory of Food Chemistry, Department of Chemistry, School of Sciences, National and Kapodistrian University of Athens, Athens, Greece*

ARNABOLDI SARA • *Department of Food Safety, Istituto Zooprofilattico della Lombardia e dell'Emilia Romagna, Brescia, Italy; Department of Veterinary Science, Università degli Studi di Parma, Parma, Italy*

DONG JOO SEO • *Department of Food and Nutrition, College of Biotechnology and Natural Resources, Chung-Ang University, Anseong, Republic of Korea*

VERÓNICA SILVA • *International Iberian Nanotechnology Laboratory (INL), Braga, Portugal; University of Minho, Braga, Portugal*

MARTA H. TANIWAKI • *Institute of Food Technology (ITAL), Campinas, Brazil*

GABRIELA LOPEZ VELASCO • *Food Safety, Neogen Corporation, Lansing, MI, USA*

MOHIT S. VERMA • *Department of Agricultural and Biological Engineering, Purdue University, West Lafayette, IN, USA; Birck Nanotechnology Center, Purdue University, West Lafayette, IN, USA; Weldon School of Biomedical Engineering, Purdue University, West Lafayette, IN, USA*

FILIPELLO VIRGINIA • *Department of Food Safety, Istituto Zooprofilattico della Lombardia e dell'Emilia Romagna, Brescia, Italy*

RAJESHWARI VITTAL • *Department of Food Safety and Biotechnology, Nitte University Centre for Science Education and Research, Nitte (Deemed to be University), Mangalore, Karnataka, India*

JIANGSHAN WANG • *Department of Agricultural and Biological Engineering, Purdue University, West Lafayette, IN, USA; Birck Nanotechnology Center, Purdue University, West Lafayette, IN, USA*

SHIYAO WANG • *PocNAT Limited, Hong Kong, China; Department of Health Technology and Informatics, The Hong Kong Polytechnic University, Hong Kong, China; Department of Biomedical Engineering, The Hong Kong Polytechnic University, Hong Kong, China*

STRYER WONG • *PocNAT Limited, Hong Kong, China*

JACKSON CHIT SHING WOO • *SGS Hong Kong Limited, Hong Kong, China*

WATARU YAMAZAKI • *Center for Southeast Asian Studies, Kyoto University, Kyoto, Japan*

QIANRU YANG • *Human Foods Program, U.S. Food and Drug Administration, Laurel, MD, USA*

DASEUL YEO • *Department of Food and Nutrition, College of Biotechnology and Natural Resources, Chung-Ang University, Anseong, Republic of Korea; Department of Food and Nutrition, College of Biotechnology and Natural Resources, Chung-Ang University, Seoul, Republic of Korea*

NADINE YERAMIAN • *Microbiology Division, Faculty of Sciences, University of Burgos, Burgos, Spain; Research Centre for Emerging Pathogens and Global Health, University of Burgos, Burgos, Spain*

SHEA PING YIP • *PocNAT Limited, Hong Kong, China; Department of Health Technology and Informatics, The Hong Kong Polytechnic University, Hong Kong, China*

Shenia R. Young • *Center for Veterinary Medicine, U.S. Food and Drug Administration, Laurel, MD, USA*
Jack Yiu Tat Yuen • *SGS Hong Kong Limited, Hong Kong, China*
Guodong Zhang • *Human Foods Program, U.S. Food and Drug Administration, College Park, MD, USA; Office of Laboratory Operations and Applied Science, Human Foods Program, U.S. Food and Drug Administration, College Park, MD, USA*
Kotsiri Zoi • *Department of Public Health, Medical School, University of Patras, Patras, Greece*

Part I

Introduction

Chapter 1

Loop-Mediated Isothermal Amplification (LAMP): A Simple Method for the Detection of Foodborne Pathogens

Kezia Kozel, Sara Healy, and Marwa M. Hassan

Abstract

Loop-mediated isothermal amplification (LAMP) is a molecular diagnostic tool that has been exponentially implemented in the diagnostics field, especially during and after the SARS-CoV-2 pandemic. LAMP enables exponential amplification of any nucleic acid targets of interest at a set temperature facilitating the use of simple heating devices. Due to the nature of how the nucleotide sequence is amplified, LAMP can produce a positive signal in under 30 minutes and exhibits increased sensitivity compared to the standard DNA amplification techniques based on Polymerase Chain Reaction (PCR). This includes techniques that are commonly used in diagnostics such as real-time quantitative (qPCR) and digital droplet PCR.

Here we summarize the different methodologies and technical considerations for setting a LAMP diagnostic assay for foodborne pathogen detection. The methods can be utilized to amplify a DNA or RNA target with minimal equipment and reagents, with the view to highlighting its application for detecting foodborne pathogens in poor resource settings.

Key words Loop-mediated isothermal amplification (LAMP), DNA, Point-of-care detection, Foodborne pathogens

1 Introduction

Loop-mediated isothermal amplification (LAMP) is an alternative molecular diagnostic option for amplifying DNA or RNA targets. LAMP was originally published at the start of the millennium [1] taking advantage of the natural activity found in *Bst* polymerase, originally isolated from the bacterium *Geobacillus stearothermophilus*. *Bst* polymerase has an optimal activity at 65 °C, exhibiting strong displacement activity enabling the displacement of DNA downstream during amplification, and lacks 3′-5′ exonuclease activity allowing for faster DNA synthesis [2]. As a result, the LAMP reaction can amplify a small amount of DNA template to 10^9 copies in under an hour [1], with initial detection being seen within 15 min of the reaction starting [3]. For amplification of

Alejandro Garrido-Maestu and Alexandre Lamas (eds.), *Foodborne Pathogens: Loop-mediated Isothermal Amplification Protocols*, Methods and Protocols in Food Science, https://doi.org/10.1007/978-1-0716-4526-0_1,

RNA targets, reverse transcriptase can be added to the reaction to enable the conversion to cDNA first before the amplification takes place and this can be all in a one step, if needed [1].

While the activity of *Bst* polymerase is a key element to the efficiency of the LAMP assay, the other is arguably primer design. The primer design for LAMP reactions enables greater sensitivity and specificity due to the use of two to three primer sets that can be applied, instead of just one. All LAMP reactions require a forward and reverse primer set (F3 and B3), which sit just outside of the forward and reverse inner primer sets (FIP and BIP) on the target template sequence. The F3 and B3 primers act to displace amplified strand initiated by the FIP and BIP primers. As shown in Fig. 1, the design of FIP and BIP is slightly different to normal primers. Both primer sets contain two main sections, a section that is complementary to the target template (FIP, F2; BIP, B2). This section allows the extension of the DNA template from both the 3′ of the target and then the 5′ end of the amplified copy of the target. The second section is complementary to the DNA product formed by the F1c and B1c sections. Dependent on publication the F1c-F2 and B2-B1c sections will also be separated by a quadruple thymidine linker or TTTT. This linker is not always seen in FIP and BIP primer design, although it has been shown that its presence can be advantageous improving the sensitivity of the LAMP assay [4]. As the template is amplified, the B1c and F1c sections self-anneal to the complementary sequences F1 and B1 within the same amplicon to form the classic LAMP dumbbell structure. On this dumbbell structure, you can have a second site of amplification, which is recognized by the third primer set, the loop primers that are designated loop forward (LF) and loop backward (LB), and FIP and BIP [1].

Traditionally the detection of amplified DNA from a LAMP assay can be summarized into three main techniques. The first amplification detection approach is observed turbidity due to magnesium pyrophosphate formation; it is a simple route to visualize DNA amplification [5, 6]. Fluorescent-based LAMP assays provide a real-time quantitative approach to DNA template detection, utilizing detection options such as DNA intercalating dyes such as SYBR green and calcein [6], or fluorescent probes (molecular beacons or TaqMan Probes) to instantly signal the presence of DNA as it is amplified. The last signal detection approach is colorimetric, which visualizes the natural decrease in pH as DNA builds in the reaction using a pH-sensitive dye such as phenol red or neutral red [6].

LAMP is advantageous in being adaptable to new technologies and protocols to improve its application in a point-of-care setting. Some brief examples utilized to detect foodborne pathogens in point-of-care settings include the synergistic application of LAMP

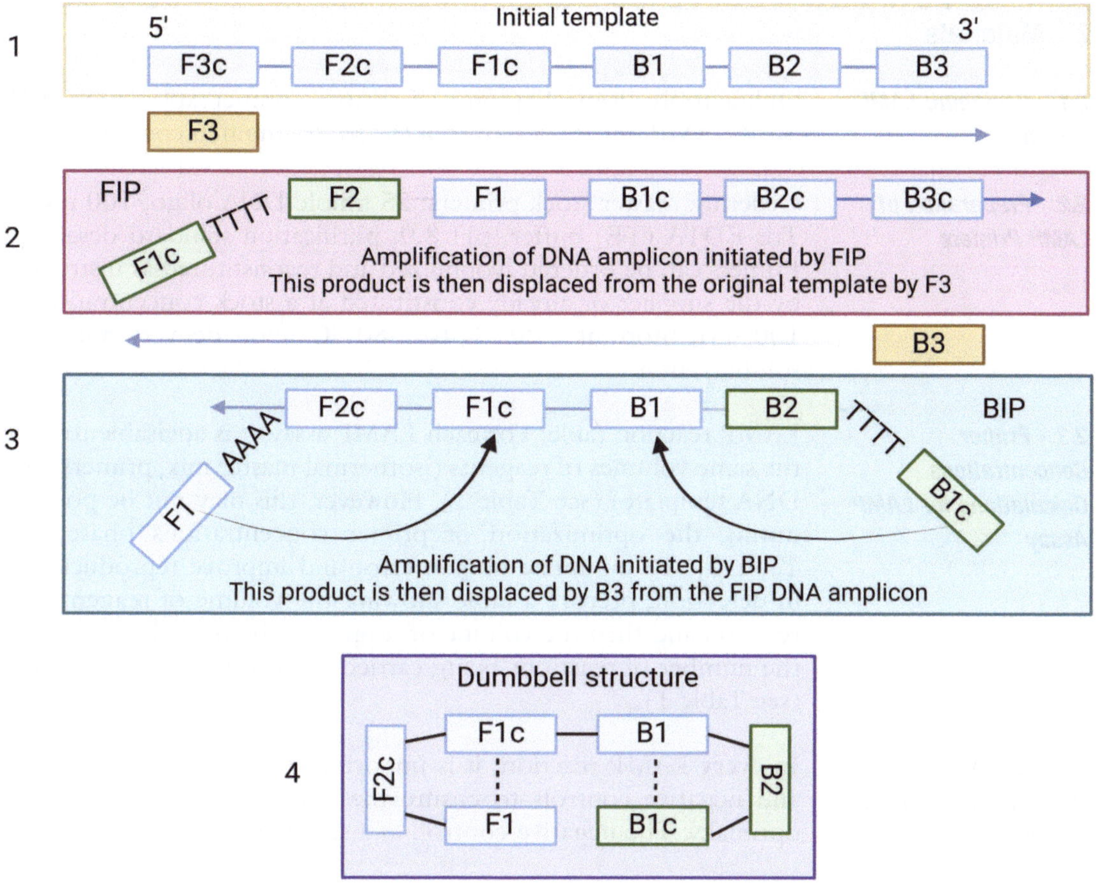

Fig. 1 Illustration of 5′-3′ DNA amplification and creation of dumbbell structure created during LAMP. This figure highlights how the inner and outer primers work synergistically to create the classic dumbbell structure seen in LAMP reactions, which then aids in further amplification, including further interaction with inner primers and loop primers. Briefly, FIP anneals to the F2c site and facilitates extension by *Bst* polymerase. The template created by this extension is then recognized by BIP through the B2 site, which then facilitates amplification of DNA back toward the 5′ end of the original target. Primers F3 and B3 act to dissociate FIP from the original DNA strand and BIP from FIP. The product from BIP annealing to the DNA product synthesized because of FIP, and DNA extension leads to the classic dumbbell sequence. The dumbbell structure is formed because F1 and F1c and B1c and B1 are complementary to each other and naturally form a hairpin. Note for ease of understanding this illustration only shows the LAMP reaction in a 5′-3′ orientation, but the reaction occurs on both strands of DNA

to lateral flow application, as has been shown for foodborne bacteria such as *Listeria monocytogenes* examples [7] and incorporation into microfluidic chips [8] to detect a variety of foodborne bacteria (*Escherichia coli* O157:H7, *Salmonella* spp., *Staphylococcus aureus*, and *Vibrio parahaemolyticus*).

This chapter will provide an initial guide to developing a LAMP assay using a fluorescent and colorimetric-based master mix to detect a DNA template of interest.

2 Materials

2.1 Designing LAMP Primers

Online tools: PrimerExplorer V5 (https://primerexplorer.jp/) [9], NEB LAMP primer design tool (https://lamp.neb.com/) [10].

2.2 Preparation of LAMP Primers

Ordering master stock primers: 25 nmole DNA oligo, 100 uM in Tris-EDTA (TE) buffer, pH 8.0, purification standard desalting. Primers can be ordered lyophilized and reconstituted as instructed by the supplier or already constituted at a stock concentration of 100 μM. Store at −20 °C to −80 °C upon delivery and after solubilization.

2.3 Primer Concentrations Calculations for LAMP Assay

LAMP reaction table: For each LAMP assay, it is advisable to have the same volumes of reagents (isothermal master mix, primers, and DNA template) (see Table 1). However, this may not be possible during the optimization of primer concentrations phase (see Table 2). To minimize human error and improve reproducibility in detection, prepare a table showing the volume of reagents per reaction and then the volume of reagents subsequently needed for the number of reactions being carried out during that experiment (see Table 1).

2.4 Establish a Positive and Negative Control

In every LAMP reaction, it is important to have internal positive and negative controls to ensure the LAMP assay is functioning optimally. The negative control, as a standard, is the nuclease-free

Table 1
Example setup of eight LAMP reactions using OptiGene Gsp SSD2.0 Isothermal Master Mix

Fluorescent LAMP Reagent	Reaction concentration		Reaction volume (rxn)	
	Start conc.	End conc.	1 rxn (μL)	8 rxn (μL)
Isothermal master mix	2×	1×	15	120
F3	10 μM	0.2 μM	0.5	4
B3	10 μM	0.2 μM	0.5	4
FIP	50 μM	2 μM	1	8
BIP	50 μM	2 μM	1	8
FLP	50 μM	1 μM	0.5	4
BLP	50 μM	1 μM	0.5	4
DNA	NA	NA	1	8
Water	NA	NA	5	40
Total			25	200

Table 2
Example of volumes required for each 50 μM primer stock per reaction when optimizing using OptiGene Gsp SSD2.0 Isothermal Master Mix

Optimizing internal and loop primers 0.4 μM–2.0 μM				
End conc. (μM)	**FIP (μL)**	**BIP (μL)**	**FLP (μL)**	**BLP (μL)**
0.4			0.2	0.2
0.6			0.3	0.3
0.8	0.4	0.4	0.4	0.4
1	0.5	0.5	0.5	0.5
1.2	0.6	0.6		
1.4	0.7	0.7		
1.6	0.8	0.8		
1.8	0.9	0.9		
2	1	1		

water aliquot that is used in your preparations. The positive control should be a sample that is known to be positive for the template DNA being detected or a synthetic sequence of DNA that is identical to the template being detected (gBlock gene fragments are often used in this case) (*see* **Note 3**). It is important to consider testing your samples matrix or elution buffer by spiking your positive control DNA as a secondary positive control to confirm the lack of amplification inhibition.

2.5 LAMP Reaction

2.5.1 Preparation of Fluorescent LAMP Reaction

Standard fluorescent LAMP reaction: Defrost the following on ice. 10 μM F3 and B3 primer stock, 50 μM FIP, BIP, FLP, and BLP primer stock, OptiGene Gsp SSD 2.0 Isothermal Master Mix, Molecular Grade Water, Target DNA. NEB also offers an isothermal master mix that can be used.

2.5.2 Colorimetric Detection of LAMP Reaction

Standard colorimetric LAMP reaction: Defrost the following on ice. 10× primer master mix (16 μM FIP and BIP, 2 μM F3 and B3, 4 μM LPF and LPB), WarmStart Colorimetric LAMP 2x Master Mix with UDG, molecular grade water, target DNA. Optigene also offers an isothermal master mix that can be used.

2.6 Optimization of LAMP

Additional reagents for optimization: 5 M betaine, magnesium sulfate ($MgSO_4$), 10 mM deoxynucleotide triphosphates (dNTPs)

Buffer: 50× Tris Acetate (400 mM), EDTA (10 mM) (TAE), buffer or 10× Tris/Borate/EDTA (TBE) buffer. Buffers are diluted to 1× in distilled or deionized water prior to use.

1.5% Agarose gel: Molecular grade agarose powder (0.75 g), 50 mL of 1× TAE or 1× TBE buffer, 10,000× SYBR safe.

Molecular marker: 6× gel loading dye, 50 or 100 bp ladder marker, 0.5 U per well.

3 Methods

3.1 Using NEB LAMP Primer Design Tool

1. Utilizing the NEB LAMP primer design tool enter in the FASTA sequence of interest by pasting it into the "Input source sequence" on the NEB LAMP primer design tool. The sequence will need to be within the range of 200–2000 base pairs long.
2. Select the "process text" option to generate a parsed sequence and then specify a relevant target name.
3. Once completed select specific parameter selection to automatically assign primers based on %GC, and then press continue.
4. In the next window, select if any of the primers need to be in a fixed specific place, and then select "generate primers" to identify if any primers can be designed against the DNA or RNA template provided. For clarification on specific template attributes and how to select the best primer sets (*see* **Note 3**).

3.2 Resolubilizing and Aliquoting Primers

Where possible primers should be handled in a PCR hood or a DNA clear zone (*see* **Note 1**). It should also be noted that primers can be ordered in 1x Tris-EDTA (TE); however if they have been ordered lyophilized (dry), they will need to be reconstituted as described below.

1. To reconstitute primers, spin down primers in a mini centrifuge, and then resolubilize primers to give 100 μM concentration using Tris-EDTA (TE) buffer pH 8.0. The volume required to make a 100 μM solution will be provided by the company that made the primers.
2. Lightly vortex, and mini centrifuge again to remove any solution from the lid of the tube.
3. For optimizing fluorescent LAMP assays, it is easier to have primer aliquots at 10 μM stocks. Clearly label microfuge tubes that will hold 10 μM aliquots of the 100 μM master stock.
4. Add 90 μL of 1× TE buffer (pH 8.0) to each microfuge tube.
5. Lightly vortex each 100 μM primer stock and then mini centrifuge prior to taking 10 μL of each and resuspend in respective labeled tube. Store at −20 °C to −80 °C.

3.3 LAMP Reaction

3.3.1 Fluorescent LAMP Assays

1. Using the standard OptiGene isothermal master mix reaction, demonstrated in Table 1, place 40 μL of molecular grade water into a 1.5 mL microfuge tube.

2. Vortex and then briefly mini centrifuge each primer aliquot and pipette 4 μL of the 10 μM F3 and B3 primers, 8 μL of the 50 μM FIP and BIP primers, and 4 μL of the 50 μM FLP and BLP primers.
3. Gently vortex the isothermal master mix and then add 120 μL to the reaction master mix. Close the microfuge tube and briefly set back done on ice.
4. Take a set of LAMP reaction tubes, and place them on ice, gently vortex the reaction master mix, and lightly mini centrifuge to remove any solution from the top of the lid.
5. Place 24 μL of the isothermal reaction master mix into each reaction tube, gently pipetting up and down between each 24 μL being taken to ensure homogenous transfer of reagents.
6. Keeping the reactions on ice, gently close the lid on each reaction tube, but do not close completely. Move the reaction tubes containing the master mix to a separate area where DNA can be handled.
7. Change lab coats to limit any DNA contamination to future LAMP setups.
8. In an area suitable for handling DNA, add 1 μL of DNA sample to each reaction tube except the seventh and the eighth well. In the seventh well, place a positive control (a DNA sample known to be positive for the DNA template amplified), and in the eighth well, place 1 μL of water that will act as a negative control. Note that more than 1 μL of DNA can be added; if necessary, the amount of water in the reaction can be adjusted to compensate for this. The volume of water in the negative control should be the same as the amount of DNA template added.
9. Place reactions in a LAMP reader (OptiGene, Genie II), incubate the reaction at 65 °C for 30 min. After 30 min, a melt curve can be programmed (suggested 98–80 °C, 0.05 °C/s), which will also inactivate the *Bst* polymerase.

3.3.2 Colorimetric LAMP Assays

1. For a standard colorimetric LAMP reaction setup using the New England Biolabs Warmstart colorimetric reaction master mix, it is advised to make a master mix of the primers prior to setting up the reaction. To do this in a microfuge tube pipette 56 μL of 1× TE buffer (pH 8.0). Then pipette the following volumes of master stock primers (100 μM) into the water. Pipette 16 μL of 100 μM FIP and BIP, 2 μL of 100 μM F3 and B3 master stock, and 4 μL of 100 μM LPF and LPB master stock. After all primer sets have been added, vortex gently and rest on ice.
2. For preparation of the colorimetric reaction, prepare a master mix for eight reactions. To do this take 72 μL of molecular grade water and add it to a 1.5 mL microfuge tube.

3. Gently vortex and pipette 2.5 μL of the 10× LAMP primer mix previously prepared in **step 1**, and then gently vortex and pipette 12.5 μL of the colorimetric LAMP master mix.
4. Once the LAMP reaction is made up, rest it on ice and set up the reaction tubes. Gently disturbing the colorimetric LAMP assay master mix by pipetting up and down, take 24 μL of the reaction mixture and add it to each tube. Then gently close each reaction tube lid taking care not to seal them completely.
5. Place the reaction tubes on ice and move them to an area where DNA can be handled. As with the fluorescent LAMP protocol, change lab coats to limit any DNA contamination and add 1 uL of DNA sample to each reaction tube except the seventh and the eighth well. In the seventh well, place a positive control (a DNA sample known to be positive for the DNA template amplified), and in the eighth well, place 1 uL of water that will act as a negative control.
6. It is then advised to incubate the tubes in a PCR machine or a heating block for 30 min at 65 °C.
7. After 30 min remove the samples from the heat and check to see if the samples have changed color. If the color change is still ongoing, samples can be heated for a further 15–30 min if needed. The color may slightly change while the tubes are cooling.
8. If there is uncertainty about color change, LAMP products can be run on a 1.5% gel; however this should be done with great caution as the reaction tube may contain highly concentrated DNA target amplicon. As such it should not be handled in any room within which the LAMP assay is set up to avoid contamination.

3.3.3 Agarose Gel with 1× TAE Buffer

1. Weigh out 0.75 g of molecular grade agarose powder and add 50 mL of 1× TAE buffer to a 150 mL glass bottle.
2. Wearing additional PPE for hot solutions (eye safety glasses and heat-resistant gloves), microwave the solution for 30 s with the lid lightly screwed on.
3. After 30 s swirl the solution and then microwave for another 30 s, stopping when the solution starts to boil.
4. Remove molten agarose gel and leave to cool slightly. When the agarose is still warm, but it can be held without heat resistant gloves, add the gel stain (e.g., 5 μL of SYBR safe).
5. Swirl the molten agarose gel so that the stain completely dissolves into the gel and then pour it into a gel case with the appropriate number of wells. Leave to set for approximately 15 min.

3.3.4 Loading 1.5% Agarose Gel with LAMP Products

1. Lower the set 1.5% agarose into the gel tank and submerge in 1× TAE buffer.
2. Load 6 μL of DNA ladder (0.5 U) to the first and last well of the 1.5% agarose gel.
3. Pipette 5 μL beads of 6× loading dye onto parafilm, do enough for each LAMP sample to be run.
4. Set the pipette to 10 μL, pipette the 5 μL of loading dye into the first LAMP sample, and pipette up and down to directly mix the loading dye into the LAMP product. Once thoroughly mixed, load 10 μL into the gel. Do this for all samples including the negative control.
5. Run the gel at 110 volts of 30 min and visualize the gel using a UV gel dock.

4 Notes

4.1 Preparation of Laboratory Workspace

Preparation of workspace is a simple step but a key consideration when undertaking a LAMP assay. LAMP is an extremely sensitive assay that can amplify DNA from on a few copies; as such it is important to ensure that target DNA is not handled in the same space as LAMP reagents. In an ideal situation, DNA extraction, the handling of LAMP reagents, addition of the DNA to be tested, and post-LAMP products would be handled in separate rooms. This is not a reality in all laboratories, especially in resource-poor settings; as such some steps can be taken to avoid contamination. Ideally DNA should not be extracted on the same day as the LAMP assay is set up, if conducted in the same room. LAMP reagents and DNA to be tested should be handled at opposite ends of the laboratory to ensure no contamination. Additionally, it is advisable to change PPE between DNA-free and DNA-present areas and to clean work-benches down thoroughly before and after LAMP setup.

4.2 Extraction of DNA from Sample

DNA extraction is a common protocol, with specialized kits that are designed for extraction of DNA from specific sources. It is advised to source a kit that is specialized to the source that the target DNA is to be extracted from. Key equipment for most DNA extraction kits includes a heat block, vortex, and centrifuge. It is advised that extracted DNA be stored at −20 or −80 °C after extraction in the kit elution buffer if possible. The latter is highlighting as not all types of molecular analysis downstream to DNA extraction favor DNA extraction kit elution buffer, such as next-generation sequencing platforms, when this is the case, elution in molecular grade water will suffice.

4.3 Designing a Synthetic DNA Target (gBlock)

Utilizing synthetic DNA, such as gBlock gene fragment from integrated DNA technologies for assay development, can be beneficial if a natural source of DNA is hard to come by or limited. Synthetic DNA can be used during the process of optimization but can also be helpful for establishing the copy number for which the assay can detect down too. Whereas with a normal DNA template, the sensitivity of the assay is based on the concentration of DNA that an assay can detect, with synthetic DNA the actual copy number of the template can be calculated. As with all concentrated DNA, the only thing to be careful with when handling synthetic DNA is contamination of the laboratory environment. If using a synthetic template, it is important to dilute it in an area where the LAMP assay itself is not set up.

4.4 Designing LAMP Primers

There are several standard rules when designing primers for LAMP; in most cases LAMP primer design software will apply these rules but is important to be aware of them. The first is there are size limitations for the chosen target template sequence; it must be >200 base pairs in size with a maximum size of 2000 bp for PrimerExplorerV5 and the NEB Primer Design tool. It should be noted however that there are new tools being developed for extending searches for LAMP primer designs to much larger templates, including the freely available Python-based tool called LAMPrimers iQ [11]. The nucleotide composition and position of primer recognition sites are also important for the efficiency of the LAMP reaction. The ends of the primers should not be AT-rich, as this can lower the Tm value, can affect the stability of the primers annealing, and can also lead to primers forming secondary structures. It is also important to check that the primer melting temperature (Tm) values are within the same temperature range that *Bst* polymerase demonstrates optimal activity (55–65 °C). The distance between F1 and F2, B1 and B2 should be 40–60 bp in length. This is important for ensuring the mechanisms of LAMP are efficient and that the positions of these sites do not hinder the LAMP reaction.

Lastly, while not automatically placed in the FIP and BIP primer design, it is advisable to place a thymidine linker "TTTT" between F1c-F2 and B2-B1c [4] as demonstrated in Fig. 1.

4.5 Optimizing Fluorescent LAMP

Although the methodology for fluorescent LAMP describes how to perform an assay using a premade isothermal reaction master mix, there are still elements in the reaction that may need to be optimized. This includes optimizing the working concentrations for both in internal primers (FIP and BIP) and the loop primers (FLP and BLP), checking the reaction temperature is optimal, and adding reagents to make the LAMP assay more efficient.

1. Optimizing primer concentrations

The inner primer concentration range for OptiGene Gsp SSD2.0 Isothermal Master Mix is 0.8–2 μM, while the recommended loop primer concentration range is 0.4–1.0 μM. It is advised to test the concentration of one primer set at a time using the normal reaction setup described above. For example, for inner primers, FIP and BIP, a concentration range of 0.8, 1.0, 1.2, 1.4, 1.6, 1.8, and 2.0 μM could be tested. For guidance on final volumes to use in a LAMP reaction using 50 μM primer stocks, see Table 2.

2. Optimizing the LAMP reaction temperature

The optimal activity of *Bst* polymerase is seen between 55 and 65 °C. While most LAMP assays are optimal at the higher end of this range, it is advised to optimize any new assay testing temperatures between 60 and 65 °C.

3. Optimizing reaction setup with additional reagents

When utilizing LAMP for the detection of foodborne pathogens, it is important to consider natural inhibitors found in food products that might interfere with the assay. Examples of such inhibitors include positively charged (cationic) polysaccharides, secondary metabolites of phenolic compounds, and calcium ions [12].

There are several organic solutions that can be added to a LAMP reaction to compensate for any inhibitor interference and prevent the creation of secondary structures as well as nonspecific amplification. Betaine is sometimes added to reactions to improve the consistency of the reaction, especially when increased sensitivity and specificity is needed. It is advised to try optimizing the reaction with 2.5–5 μL of 5 M betaine. Dimethylsulfoxide (DMSO) is also a common additive to LAMP reactions, but the amount of DMSO should be optimized, with a suggested range of 5–10% of the reaction [13]. Overall, it is important to be aware of key inhibitors present in the food source being tested and to allow for this when undergoing LAMP assay design and optimization.

For particularly low DNA template samples, another reaction component that can be added is guanidine hydrochloride, which denatures double-stranded DNA, allowing easier access for *Bst* polymerase. This component can have inhibitory effects on the reaction as well, so it is important to test a range of concentrations, typically between 20 and 50 mM [14].

Overall, it is important to be cautious when adding supplements to LAMP reactions, and to optimize the assay systematically, using a sensible range of concentrations for each supplement, as these additional elements to the reaction may not always have a positive effect on the sensitivity and specificity of the reaction [15].

4.6 Avoiding Cross Contamination

Another important element to consider for LAMP reactions is uracil-DNA glycosylase (UDG). This enzyme specifically digests deoxyuridine triphosphate (dUTP), which is sometimes incorporated into PCR products as a type of contamination fail safe for downstream applications. This is more commonly used in laboratories where a lot of molecular diagnostics are being performed. If LAMP is being conducted in a laboratory environment where a lot of molecular diagnostic assays are set up, and dUTP is used, UDG should be considered as an additional element of the LAMP assay being optimized to ensure false positives, or heightened positive reactions are not seen from carryover contamination [16, 17].

4.7 Visualizing LAMP Products on a 1.5% Agarose Gel

It is important to note that successful amplification of a DNA amplicon through LAMP can lead to highly concentrated amplicon ($< 10^9$ copies). Opening any tube that has contained a LAMP reaction immediately risks contaminating the surrounding laboratory environment with aerosolized DNA amplicon. As such this product should only be run on a gel, if necessary, for example, confirmation of colorimetric assay being successful during optimization. If this needs to be done, the LAMP tubes should be spun down on a mini centrifuge and opened in a separate room to where the LAMP assay is set up to remove any chance of contaminating ongoing assays.

References

1. Notomi T, Okayama H, Masubuchi H, Yonekawa T, Watanabe K, Amino N, Hase T (2000) Loop-mediated isothermal amplification of DNA. Nucleic Acids Res 28(12):e63–e63
2. Oscorbin I, Filipenko M (2023) *Bst* polymerase – a humble relative of Taq polymerase. Computational and structural. Biotechnol J 21:4519
3. Hassan MM, Grist LF, Poirier AC, La Ragione RM (2022) JMM profile: Loop-mediated isothermal amplification (LAMP): for the rapid detection of nucleic acid targets in resource-limited settings. J Med Microbiol 71(5): 001522
4. Lamas A, Azinheiro S, Roumani F, Prado M, Garrido-Maestu A (2023) Evaluation of the effect of outer primer structure, and inner primer linker sequences, in the performance of loop-mediated isothermal amplification. Talanta 260:124642
5. Mori Y, Nagamine K, Tomita N, Notomi T (2001) Detection of Loop-mediated isothermal amplification reaction by turbidity derived from magnesium pyrophosphate formation. Biochem Biophys Res Commun 289(1):150–154
6. Fischbach J, Xander NC, Frohme M, Glökler JF (2015) Shining a light on LAMP assays – a comparison of LAMP visualization methods including the novel use of berberine. BioTechniques 58(4):189–194. https://doi.org/10.2144/000114275
7. Yu Z, Shao Y, Dong Y, Wen T, Chen Z, Zhang Y, Cheng F, Wang Z, Tu J, Qi K (2024) LAMP combined with *Pyrococcus furiosus* argonaute for the ultrasensitive and highly specific point-of-care test platform for listeria monocytogenes detection. LWT 207:116640
8. Zhang M, Liu J, Shen Z, Liu Y, Song Y, Liang Y, Li Z, Nie L, Fang Y, Zhao Y (2021) A newly developed paper embedded microchip based on LAMP for rapid multiple detections of foodborne pathogens. BMC Microbiol 21(1):197. https://doi.org/10.1186/s12866-021-02223-0

9. PrimerExplorer V5 (2024). https://primerexplorer.jp/lampv5e/index.html. Accessed 26 Nov 2024
10. NEB LAMP Primer Design Tool Version 1.4.2 (2024). https://lamp.neb.com/#!/. Accessed 26 Nov 2024
11. Akhmetzianova LU, Davletkulov TM, Sakhabutdinova AR, Chemeris AV, Gubaydullin IM, Garafutdinov RR (2024) LAMPrimers iQ: new primer design software for Loop-mediated isothermal amplification (LAMP). Anal Biochem 684:115376
12. Moon Y-J, Lee S-Y, Oh S-W (2022) A review of isothermal amplification methods and food-origin inhibitors against detecting food-borne pathogens. Food Secur 11(3):322
13. Wang DG, Brewster JD, Paul M, Tomasula PM (2015) Two methods for increased specificity and sensitivity in Loop-mediated isothermal amplification. Molecules 20(4):6048–6059. https://doi.org/10.3390/molecules20046048
14. Zhang Y, Ren G, Buss J, Barry AJ, Patton GC, Tanner NA (2020) Enhancing colorimetric Loop-mediated isothermal amplification speed and sensitivity with guanidine chloride. BioTechniques 69(3):178–185. https://doi.org/10.2144/btn-2020-0078
15. Costa-Ribeiro A, Lamas A, Garrido-Maestu A (2024) Evaluating commercial Loop-mediated isothermal amplification master mixes for enhanced detection of foodborne pathogens. Food Secur 13(11):1635
16. Hsieh K, Mage PL, Csordas AT, Eisenstein M, Soh HT (2014) Simultaneous elimination of carryover contamination and detection of DNA with uracil-DNA-glycosylase-supplemented Loop-mediated isothermal amplification (UDG-LAMP). Chem Commun 50(28): 3747–3749
17. Wu F, Lu C, Hu W, Guo X, Chen J, Luo Z (2023) Rapid visual detection of *Vibrio parahaemolyticus* by combining LAMP-CRISPR/Cas12b with heat-labile uracil-DNA glycosylase to eliminate carry-over contamination. J Zhejiang Univ Sci B 24(8):749–754

Part II

Regulated Pathogens

Chapter 2

Rapid Detection of *Salmonella* in Frozen Fish and Dry Dog Food

Qianru Yang, Kelly J. Domesle, Shenia R. Young, Ryan C. McDonald, Guodong Zhang, and Beilei Ge

Abstract

Many *Salmonella* LAMP assays have been developed and applied toward the rapid, reliable, and robust detection of *Salmonella* in foods. The *Salmonella* LAMP assay detailed in this protocol specifically targets the invasion gene (*invA*) and has been validated in 34 human food and animal food matrices to date. Using frozen fish and dry dog food as representative matrices, samples are enriched and processed for nucleic acid extraction before the LAMP analysis. The LAMP assay can be performed on a small fluorometer or a standard qPCR instrument. Data interpretation includes evaluation of controls, positive/negative sample calls, and suggested repeats for inconclusive results. This *Salmonella* LAMP protocol is an invaluable tool for analyzing foods for *Salmonella* contamination.

Key words Animal food, Detection, Dry dog food, Food, Frozen fish, Human food, Loop-mediated isothermal amplification, *Salmonella*

1 Introduction

Loop-mediated isothermal amplification (LAMP) is an isothermal nucleic acid amplification test (NAAT) invented by Japanese scientists Notomi et al. in 2000 [1]. LAMP has grown exponentially in both popularity and applicability over the last two decades, establishing itself as a powerful alternative to polymerase chain reaction (PCR) for the detection of microorganisms and associated genetic markers in various fields [2–5]. The LAMP amplification strategy is based on the initial formation of a single-stranded DNA shaped like a dumbbell with loops at both ends, which serves as the starting material for efficient LAMP cycling [1]. LAMP offers many advantages over PCR and other NAATs, including isothermal amplification, high amplification efficiency, versatile detection platforms, and a high tolerance to matrix effects [6–8].

Alejandro Garrido-Maestu and Alexandre Lamas (eds.), *Foodborne Pathogens: Loop-mediated Isothermal Amplification Protocols*, Methods and Protocols in Food Science, https://doi.org/10.1007/978-1-0716-4526-0_2,

Many LAMP assays have been developed and used for *Salmonella* detection in foods as summarized in a comprehensive review [9]. The *Salmonella* LAMP assay detailed in this protocol specifically targets the invasion gene *invA* (GenBank accession number M90846) [10] and is rapid, reliable, and robust in a variety of human and animal food matrices [8, 11–16]. The limits of detection of this assay when tested using different *Salmonella* serovars in pure culture range from 1.3 to 28 cells per reaction [15]. In food matrices, the assay can detect one colony-forming unit (CFU) in 25 g test portions with overnight enrichment [15]. This *Salmonella* LAMP assay has been validated in multiple food categories (34 matrices to date from 9 ISO categories [17]; Table 1) in single-laboratory validation studies [15, 18, 1000] (also an in-press publication in *Food Microbiology*) and validated in dry dog food through a multi-laboratory validation study [19]. As of 2022, this *Salmonella* LAMP assay has been adopted by the National Antimicrobial Resistance Monitoring System (NARMS) as the uniform *Salmonella* screening methods in retail meat testing [20].

Using frozen fish and dry dog food as representative human and animal food matrices, we describe here the complete LAMP protocol consisting of sample preparation, DNA extraction, and LAMP analysis (reaction assembly, run, and data interpretation) (Fig. 1). The LAMP assay is performed on two platforms: frozen fish samples on a standard real-time PCR (qPCR) instrument (7500 Fast; Applied Biosystems, part of Thermo Fisher Scientific, Waltham, MA, USA) (Figs. 2 and 5) and dry dog food samples on a portable fluorometer (Genie II; OptiGene, Ltd., West Sussex, UK) (Figs. 3 and 4).

2 Materials

Prepare all media and reagents with distilled water and store at conditions instructed by the manufacturers. Prepare LAMP primer mix and reaction mix with molecular grade water. Diligently follow all waste disposal regulations.

2.1 Equipment and Supplies

1. Genie II (OptiGene Ltd.).
2. Genie Strips (OptiGene Ltd.).
3. Genie Strip Holder (OptiGene Ltd.).
4. 7500 Fast (Thermo Fisher Scientific).
5. MicroAmp Fast Optical 96-Well Reaction Plates or MicroAmp Fast 8-Tube Strips (Thermo Fisher Scientific).
6. MicroAmp Optical Adhesive Film or MicroAmp Optical 8-Cap Strips (Thermo Fisher Scientific).
7. MicroAmp 96-Well Support Base (Thermo Fisher Scientific).

Table 1
ISO food categories, types, and matrices validated by the *Salmonella* LAMP assay

ISO food category	Type	Matrix
Chocolate, bakery products, and confectionary	Dry and sugared low moisture ($a_w < 0.65$)	Cocoa powder
	Dry powdered	Cake mixes
	Low moisture	Cookies
Dried cereals, fruits, nuts, seeds, and vegetables	Flours	All-purpose flour
	Nuts and seeds	Peanut butter
	Seasonings	Ground black pepper
Eggs and egg products (derivatives)	Dry	Egg powder
	Eggs (unprocessed)	Shell eggs
	Egg products (heat processed) without additives	Liquid eggs
Environmental samples (food or feed production)	Equipment or production environment	Sealed concrete surface sampled with sponges
		Stainless-steel surface sampled with swabs
	Waters used in the manufacturing process	Recycled washing water
Fish and seafood products	Crustaceans (unprocessed)	Frozen shrimp
	Fish (unprocessed)	Frozen tilapia
	Smoked or cured, and other processed products ($a_w > 0.92$)	Smoked fish
Fresh produce and fruits	Cut ready-to-eat fruits	Cantaloupe
	Leafy greens	Romaine lettuce
	Sprouts	Alfalfa sprouts
Infant formula and infant cereals	Non-probiotic infant cereals	Infant cereals
	Non-probiotic infant formula	Whey-based formula
	Non-probiotic ingredients	Nonfat dry milk
Milk and dairy products	Pasteurized dairy products	Ice cream
	Pasteurized milk-based products	Soft cheese
	Raw milks and/or fermented/acidified milks (not heat treated)	Raw milk
Pet food and animal feed	Animal feeds (bovine, ovine, pig)	Cattle feed
		Dairy feed
		Horse feed
		Swine feed
	Animal feeds (poultry)	Poultry feed
	Dry food ($a_w \leq 0.7$)	Dry cat food
		Dry dog food
		Pig ear
	Wet food ($a_w > 0.7$)	Raw beef pet food
		Raw chicken pet food

Note: For two categories, "raw" and "processed (products)" are combined into one category. These include the "fish and seafood products" category and the "milk and dairy products" category

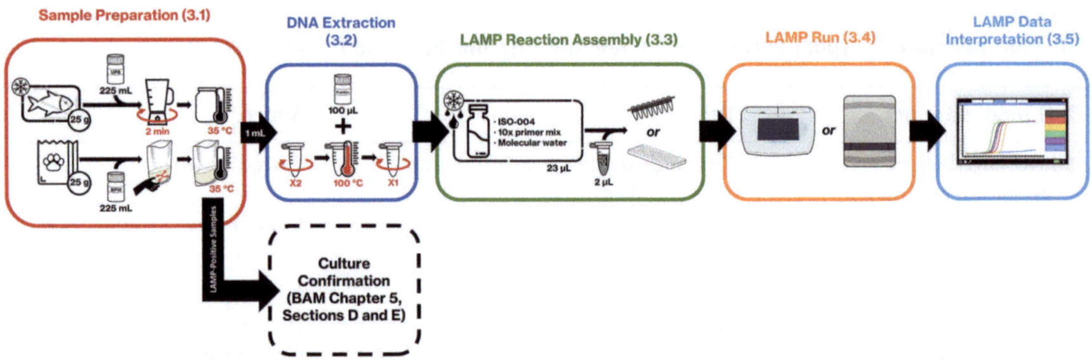

Fig. 1 A schematic diagram of the *Salmonella* LAMP workflow. Abbreviations in the order of appearance are as follows: UPB, universal preenrichment broth; BPW, buffered peptone water; BAM, the US Food and Drug Administration's *Bacteriological Analytical Manual*; LAMP, loop-mediated isothermal amplification; ISO-004, GspSSD2.0 Isothermal Mastermix. **IMPORTANT**: All LAMP-positive samples should be confirmed by culture [21]

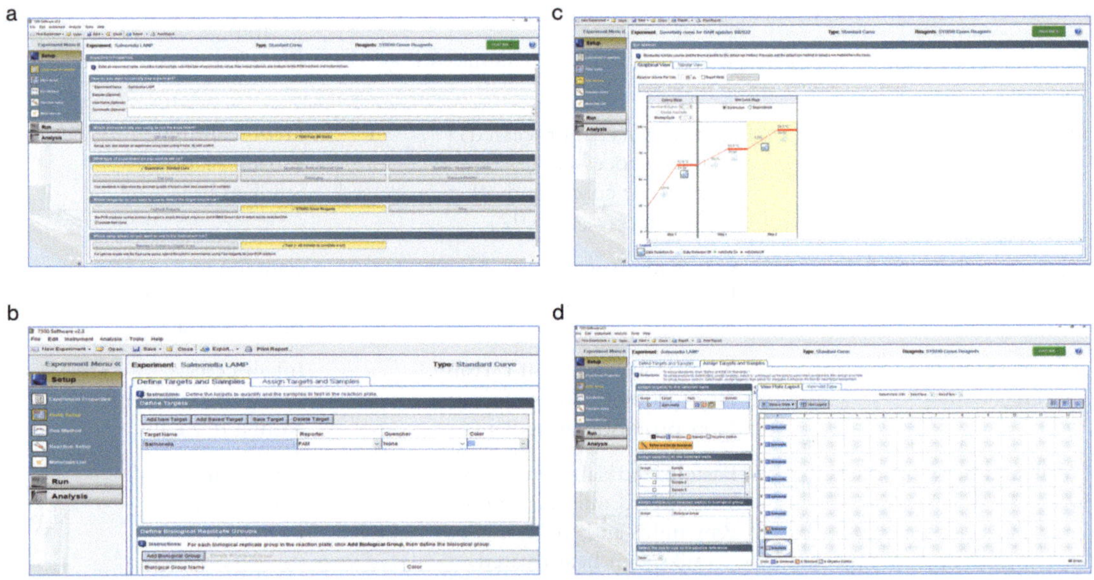

Fig. 2 Screenshots to guide 7500 Fast setup steps. Panel (**a**) shows the [Experimental Properties] tab. Panel (**b**) shows the [Plate Setup] tab to [Define Targets and Samples]. Panel (**c**) shows the [Run Method] tab for [Graphical View]. Panel (**d**) shows the [Plate Setup] tab to [Assign Targets and Samples]

8. Mini plate spinner centrifuge.
9. Pipettes (0.5–10, 2–20, 20–200, and 200–1000 μL) and tips (aerosol-resistant).
10. Vortex mixer.
11. Microcentrifuge and tubes (0.5–2 mL) (*see* **Note 1**).
12. Heat block.

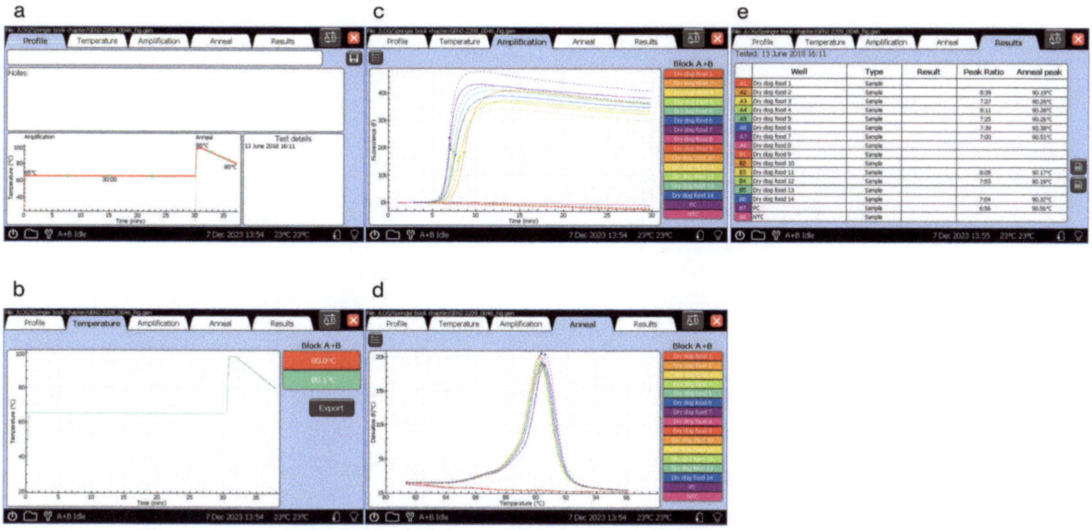

Fig. 3 Representative LAMP results displayed on the Genie II instrument panel (firmware v2.34.15). In this LAMP run, A1 to A8 and B1 to B6 are dry dog food samples 1–14. B7 and B8 are PC and NTC, respectively. PC is *S.* Typhimurium ATCC 19585 (LT2) at ca. 10^4 cells per reaction and NTC is molecular grade water. Panel (**a**) shows the [Profile] tab for the programmed temperature profile. Panel (**b**) shows the [Temperature] tab for actual temperatures in the sample wells as LAMP reaction proceeds. Panel (**c**) shows the [Amplification] tab for fluorescence readings during LAMP amplification. Panel (**d**) shows the [Anneal] tab for changes in fluorescence (derivative) during the anneal phase. Panel (**e**) shows the [Results] tab for a tabular view of the LAMP results

13. Sterile spoons or other appropriate instruments for transferring food samples.
14. Sterile knives for cutting samples.
15. Balance with weights.
16. Blender and sterile blender jars.
17. Sterile wide-mouth screw-cap jars (500 mL).
18. Sterile sample filter bags and racks.
19. Sterile culture dishes (15 × 100 mm).
20. Inoculating needles and loops.

2.2 Media and Reagents

1. Trypticase soy agar (TSA): 15 g pancreatic digest of casein, 5 g papaic digest of soybean, 5 g sodium chloride, and 15 g agar. Heat with gentle agitation to dissolve in 1 L distilled water and autoclave at 121 °C for 15 min. Final pH, 7.3 ± 0.2.
2. Trypticase soy broth (TSB): 17 g pancreatic digest of casein, 3 g papaic digest of soybean, 5 g sodium chloride, 2.5 g dipotassium phosphate, and 2.5 g dextrose. Heat with gentle agitation to dissolve in 1 L distilled water and autoclave at 121 °C for 15 min. Final pH, 7.3 ± 0.2.

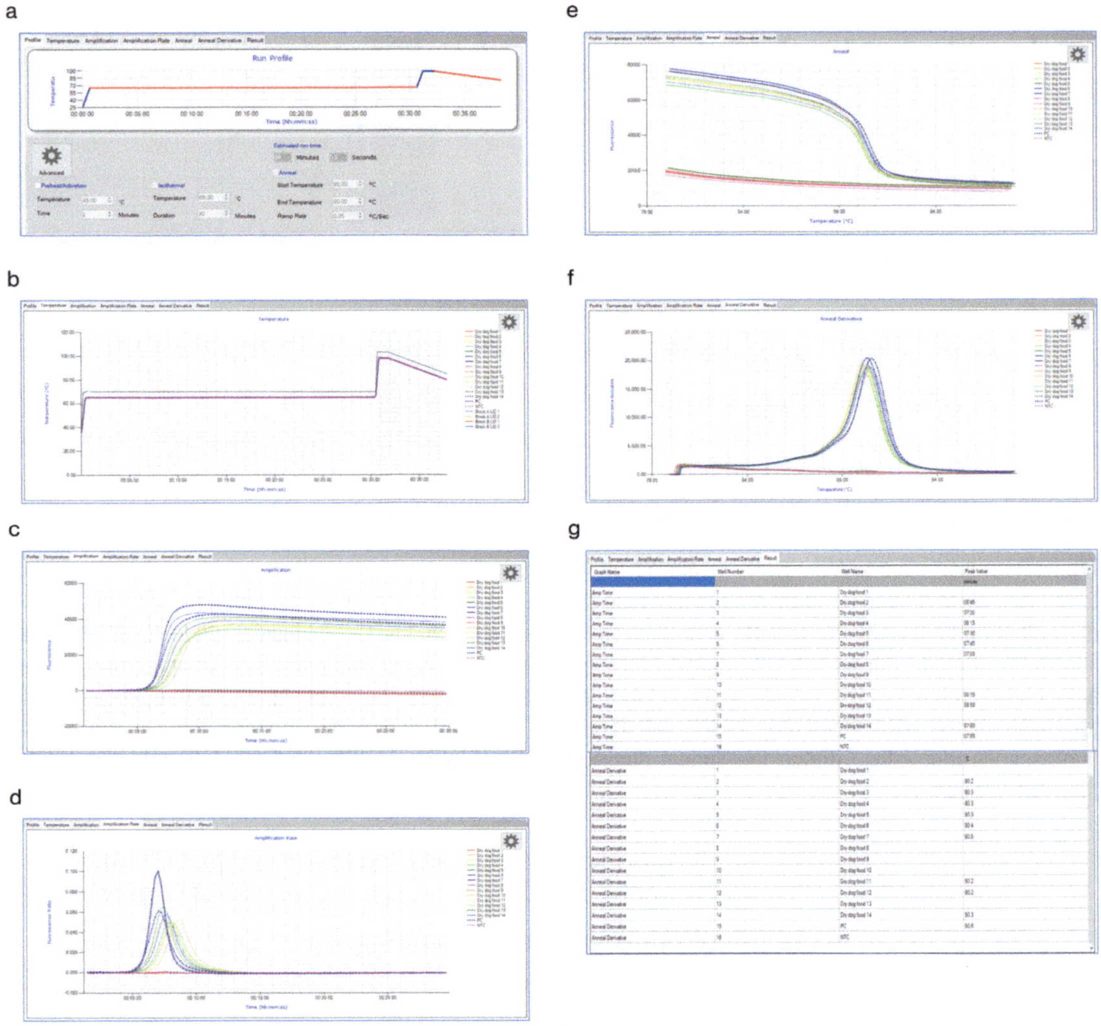

Fig. 4 Representative LAMP results viewed in the Genie Explorer software (v 2.0.7.11). This is the same run as in Fig. 3. Panel (**a**) shows the [Profile] tab for the programmed temperature profile. Panel (**b**) shows the [Temperature] tab for actual temperatures in the sample wells as LAMP reaction proceeds. Panel (**c**) shows the [Amplification] tab for fluorescence readings during LAMP amplification. Panel (**d**) shows the [Amplification Rate] tab for changes in fluorescence (fluorescence ratio) during LAMP amplification. Panel (**e**) shows the [Anneal] tab for fluorescence readings during the anneal phase. Panel (**f**) shows the [Anneal Derivative] tab for changes in fluorescence (derivative) during the anneal phase. Panel (**g**) shows the [Results] tab for a tabular view of the LAMP results

3. Peptone diluent (0.1%): 1 g peptone. Dissolve in 1 L distilled water and autoclave at 121 °C for 15 min. Final pH, 7.0 ± 0.2.
4. Universal preenrichment broth (UPB): Tryptone 5 g, proteose peptone 5 g, monopotassium phosphate 15 g, disodium hydrogen phosphate 7 g, sodium chloride 5 g, dextrose 0.5 g, magnesium sulfate 0.25 g, ferric ammonium citrate 0.1 g, sodium

pyruvate 0.2 g. Heat with gentle agitation to dissolve in 1 L distilled water and autoclave at 121 °C for 15 min. Final pH, 6.3 ± 0.2.

5. Buffered peptone water (BPW): 10 g peptone, 5 g sodium chloride, 3.5 g disodium phosphate, and 1.5 g monopotassium phosphate. Dissolve in 1 L distilled water and autoclave at 121 °C for 15 min. Final pH, 7.2 ± 0.2.
6. PrepMan Ultra Sample Preparation Reagent (Thermo Fisher Scientific).
7. Isopropanol (70%) (*see* **Note 2**).
8. DNA AWAY or equivalent (*see* **Note 3**).
9. GspSSD2.0 Isothermal Mastermix (ISO-004) (OptiGene Ltd.) (*see* **Note 4**).
10. Molecular grade water (*see* **Note 5**).
11. LAMP primers (*see* **Note 6**): Prepare stock solutions of each *Salmonella* LAMP primer (Integrated DNA Technologies, Coralville, IA or equivalent; Table 2) at 100 μM by rehydrating it with appropriate amount of molecular grade water. Mix well by vortexing for 10 s and store at −20 °C. Prepare the LAMP primer mix (10×) according to the worksheet in Table 3. Add appropriate volumes of stock solutions for each primer and molecular grade water into a microcentrifuge tube. Mix all reagents well by vortexing for 10 s. Aliquot to 500 μL per microcentrifuge tube and store at −20 °C (*see* **Note** 7).
12. LAMP controls (*see* **Note 8**). To prepare a positive control (PC), inoculate *Salmonella enterica* serovar Typhimurium ATCC 19585 (LT2) or any *Salmonella* reference strain on a nonselective agar plate (e.g., TSA) and incubate at 35 °C for 24 ± 2 h. Transfer 3–5 isolated colonies to 5 mL TSB and incubate at 35 °C for 16 ± 2 h to reach ca. 10^9 CFU/mL. Serially dilute the overnight culture in 0.1% peptone diluent to

Table 2
LAMP primers for *Salmonella* detection in foods

Primer name	Sequence (5′-3′)	Length (bp)
Sal4-F3	GAACGTGTCGCGGAAGTC	18
Sal4-B3	CGGCAATAGCGTCACCTT	18
Sal4-FIP	GCGCGGCATCCGCATCAATA- TCTGGATGGTATGCCCGG	38
Sal4-BIP	GCGAACGGCGAAGCGTACTG- TCGCACCGTCAAAGGAAC	38
Sal4-LF	TCAAATCGGCATCAATACTCATCTG	25
Sal4-LB	AAAGGGAAAGCCAGCTTTACG	21

Table 3
Worksheet for preparing the LAMP primer mix (10×) in 1 mL

Component	Volume (μL)
Sal4-F3 primer (100 μM)	10
Sal4-B3 primer (100 μM)	10
Sal4-FIP primer (100 μM)	180
Sal4-BIP primer (100 μM)	180
Sal4-LF primer (100 μM)	100
Sal4-LB primer (100 μM)	100
Molecular grade water	420
Total	1000

obtain ca. 10^7 CFU/mL. Transfer 500 μL of this dilution to a microcentrifuge tube and heat at 100 °C for 10 min in a heat block. Cool to room temperature, centrifuge again at 12,000× *g* for 2 min, and transfer supernatant to a new microcentrifuge tube. Store all tubes containing control DNA extracts at −20 °C. For the no template control (NTC), molecular grade water is used (*see* **Note 9**).

3 Methods

3.1 Sample Preparation

3.1.1 Sample Preparation for Frozen Fish

1. Aseptically weigh 25 g frozen fish sample into a sterile blender jar (*see* **Note 10**).
2. Add 225 mL UPB and blend for 2 min (*see* **Note 11**).
3. Aseptically transfer homogenized mixture to a sterile, wide-mouth, screw-cap jar (500 mL), or other appropriate container. Mix well and loosen jar cap about ¼ turn. Incubate at 35 °C for 24 ± 2 h.
4. Mix well by swirling the jar containing the enriched frozen fish sample. Transfer 1 mL from supernatant to a microcentrifuge tube. Vortex briefly (*see* **Note 12**).

3.1.2 Sample Preparation for Dry Dog Food

1. Aseptically weigh 25 g dry dog food into a sterile sample filter bag. Place the bag into a large container or rack for support during incubation (*see* **Note 13**).
2. Add 225 mL BPW and mix well by swirling and briefly hand massage for 2 min. Incubate at 35 °C for 24 ± 2 h.
3. Mix well by swirling the bag containing the enriched dry dog food sample. Transfer 1 mL from the filtered side of the bag to a microcentrifuge tube. Vortex briefly (*see* **Note 14**).

3.2 DNA Extraction

1. Centrifuge the microcentrifuge tube at 900× *g* for 1 min to remove large particles and transfer supernatant to a new microcentrifuge tube (*see* **Note 1**).
2. Centrifuge the new microcentrifuge tube at 16,000× *g* for 2 min and discard supernatant.
3. Suspend the pellet in 100 μL PrepMan Ultra Sample Preparation Reagent and heat at 100 °C for 10 min in a heat block.
4. Cool to room temperature and centrifuge again at 12,000× *g* for 2 min.
5. Transfer supernatant to another new microcentrifuge tube. Store all tubes containing sample DNA extracts at 4 °C if setting up the LAMP assay on the same day; otherwise, store at −20 °C.

3.3 LAMP Reaction Assembly

1. Clean the workspace, pipettes, and plate/strip holders with 70% isopropanol and DNA AWAY (*see* **Note 15**).
2. Thaw GspSSD2.0 Isothermal Mastermix (ISO-004), 10× LAMP primer mix, molecular grade water (*see* **Note 5**), PC *Salmonella* DNA, and sample DNA extracts at room temperature.
3. Prepare the LAMP reaction mix according to the worksheet in Table 4 (*see* **Note 16**). Add appropriate volumes of ISO-004, 10× LAMP primer mix, and molecular grade water into a microcentrifuge tube. Vortex gently and then flick wrist to ensure all reagents are pooled at the bottom of the tube.
4. Place the Genie Strip or MicroAmp Fast Reaction Plate/Strip in the holder and distribute 23 μL LAMP reaction mix to each well.
5. Vortex all DNA templates and centrifuge briefly to pool liquid. In a separate clean area, add 2 μL DNA template to each well and cap tightly (*see* **Note 17**).

Table 4
Worksheet for preparing the LAMP reaction mix when running different numbers of reactions

Component	Volume (μL) for 1 reaction	Volume (μL) for 9 reactions	Volume (μL) for 18 reactions	Volume (μL) for 100 reactions
GspSSD2.0 isothermal Mastermix (ISO-004)	15	135	270	1500
Primer mix (10×)	2.5	22.5	45	250
Molecular grade water	5.5	49.5	99	550
Total	23	207	414	2300

Note: DNA template (2 μL) is to be added to each reaction tube

6. Remove the Genie Strip or MicroAmp Fast Reaction Plate/ Strip from the holder and flick wrist to ensure all reagents are pooled at the bottom of the tube. Centrifuge the MicroAmp Fast Reaction Plate briefly in a mini plate spinner as needed.
7. Load the Genie Strip into Genie II block(s) or the MicroAmp Fast Reaction Plate/Strip into the 7500 Fast plate holder, ensuring caps are secure and completely sealed before closing the instrument's lid or tray.

3.4 LAMP Run

3.4.1 LAMP Run on Genie II

1. Turn on Genie II (Mark II, GEN2–02 Firmware v2.34.15) and tap once to access the home screen.
2. Tap [LAMP & Anneal] and select [Edit] to access the run profile screen (*see* **Note 18**).
3. Enter sample names using the touchscreen keyboard and tap the [Check] icon to return to the run profile screen.
4. Click the [Start] icon in the upper right of the screen and select the block(s) containing Genie Strip(s) to start the run.
5. The run saves automatically when complete.

3.4.2 LAMP Run on 7500 Fast

1. Turn on 7500 Fast and launch 7500 software (v2.3) from a connected computer.
2. Set up a LAMP experiment template by clicking [Advanced Setup] from the home screen following **steps 3–5** and save it as "Salmonella LAMP.edt" for all future LAMP runs.
3. In the [Experimental Properties] tab (Fig. 2a), select the following parameters: (1) for instrument, "7500 Fast (96 Wells)"; (2) for type of experiment, "Quantitation – Standard Curve"; (3) for reagents, "SYBR® Green Reagents" and check [Include Melt Curve]; and (4) for ramp speed, "Fast (~ 40 minutes to complete a run)."
4. In the [Plate Setup] tab (Fig. 2b), under [Define Targets]: Enter "Salmonella" for [Target Name] and select "FAM" as the [Reporter] and "None" for [Quencher].
5. In the [Run Method] tab (Fig. 2c), select either [Graphical View] or [Tabular View] and set up the following: (1) For [Reaction Volume Per Well], enter "25" μL. (2) Delete [Holding Stage], step 2 of the [Cycling Stage], and steps 2 and 4 of the [Melt Curve Stage]. (3) For [Cycling Stage], enter "30" for [Number of Cycles] and "2" for [Starting Cycle], use [Ramp Rate] of 100%, [Hold Temperature] of 65 °C, and [Hold Time] of 1:00, and select "Collect Data on Hold." (4) For [Melt Curve Stage], select "Continuous," for step 1—enter [Ramp Rate] of 100%, [Hold Temperature] of 80 °C, [Hold Time] of 1:00, for step 2—enter [Ramp Rate] of 1.5%, [Hold Temperature] of 98 °C, [Hold Time] of 0.05, and select "Collect Data on Ramp."

6. Set up a new run and enter experiment name and sample information following **steps 7–10**.
7. Open the "Salmonella LAMP.edt" template from the home screen.
8. In the [Experimental Properties] tab (Fig. 2a), type in an experiment name.
9. In the [Plate Setup] tab, under [Define Samples] (Fig. 2d), click [Add New Sample] and type in sample names. View tab [Assign Targets and Samples] and assign targets and samples to the plate layout. **IMPORTANT**: Make sure that under [select the dye to use as the passive reference] on the bottom left, change from "ROX" to "None."
10. Save the experiment file (.eds), using the experiment name used in **step 8**.
11. Click the [Start Run] button from the top right corner of the screen (*see* **Note 19**).
12. The run saves automatically when complete.

3.5 LAMP Data Interpretation (See Note 20)

3.5.1 Genie II: LAMP Result Interpretation on the Instrument Panel (Mark II, GEN2-02 Firmware v2.34.15)

1. Turn on the Genie II instrument (*see* **Note 21**).
2. Tap the [Folder] icon in the bottom left corner of the home screen and select [log] to navigate to the run file (.gen) for the LAMP run of interest.
3. Observe the five tabs associated with each run: [Profile], [Temperature], [Amplification], [Anneal], and [Results] (Fig. 3, *see* **Note 22**).
4. Tap the [Results] tab for a tabular view of the results (Fig. 3e). There are five columns (Well, Type, Result, Peak Ratio, and Anneal Peak). The "Peak Ratio" column shows the time-to-peak values (min:sec) for each sample ("Well") and the "Anneal Peak" column shows the annealing temperatures (°C) for any amplified product in that well. **Use the recommended abbreviations: T_p for "Peak Ratio" and T_a for "Anneal Peak."**
5. Examine the control wells first. For the NTC well, the "Peak Ratio (T_p)" should be blank, while the "Anneal Peak (T_a)" can be either blank or <83 °C. The PC well should have "Peak Ratio (T_p)" less than 10 min and "Anneal Peak (T_a)" around 89 ± 2 °C (*see* **Note 23**). Runs with either or both controls not in these ranges are considered invalid and should be repeated.
6. Examine the sample wells. **All samples with "Peak Ratio (T_p)" ≤ 20 min and the correct "Anneal Peak (T_a)" (89 ± 2 °C) are considered LAMP-positive for *Salmonella*.** For samples with the correct "Anneal Peak (T_a)" (89 ± 2 °C) but "Peak Ratio (T_p)" > 20 min or blank, manually observe amplification curves and repeat LAMP on sample DNA extracts

having valid LAMP amplifications (typical sigmoid amplification curves as shown in Fig. 3c). If the same results are obtained for the repeat run, the samples are considered LAMP-positive for *Salmonella*. **IMPORTANT**: All LAMP-positive samples should be confirmed by culture [21].

7. Tap the [Results] tab, export data to a single page pdf report or a CSV file or print the results directly. These outputs can also be transferred to a computer through the USB drive or with a USB memory stick.

3.5.2 Genie II: LAMP Result Interpretation on the Genie Explorer Software (v2.0.7.11)

1. Open the Genie Explorer software (*see* **Note 24**).
2. Click the [Folder] icon on the top ribbon to bring up the [Computer] icon on the left panel, and then click the [Computer] icon to navigate to the run file (.gen) for the LAMP run of interest (*see* **Note 25**).
3. Observe the seven tabs associated with each run: [Profile], [Temperature], [Amplification], [Amplification Rate], [Anneal], [Anneal Derivative], and [Result] (Fig. 4; *see* **Note 26**).
4. Click the [Amplification Rate] tab to view a graphic display of the fluorescence ratios by time (Fig. 4d). Click the [Graph Settings] icon at the top right of the screen and adjust the "Peak Detection Threshold Ratio" from 0.020 to 0.010 (*see* **Note 27**).
5. Click the [Result] tab for a tabular view of the results (Fig. 4g). There are four columns (Graph Name, Well Number, Well Name, and Peak Value) with "Amp Time" shown in the top block and "Anneal Derivative" shown in the bottom block for each well (*see* **Note 28**). **Use the recommended abbreviations: T_p to refer to "Peak Value" for "Amp Time" and T_a to refer to "Peak Value" for "Anneal Peak."**
6. Interpret the LAMP results (evaluation of controls, positive/negative sample calls, and suggested repeats for inconclusive results) and report final LAMP results following similar steps as when using the Genie II instrument panel (see Subheading 3.5.1, **steps 4–6**) with one exception that the NTC well and other negative samples should have blank "Peak Value" for "Anneal Derivative" (T_a) as the LAMP software settings eliminate those with <83 °C results. **IMPORTANT**: All LAMP-positive samples should be confirmed by culture [21].
7. Export data as a text file or an image file or generate a Genie experiment report in pdf format from the run by clicking the [Generate report] icon.

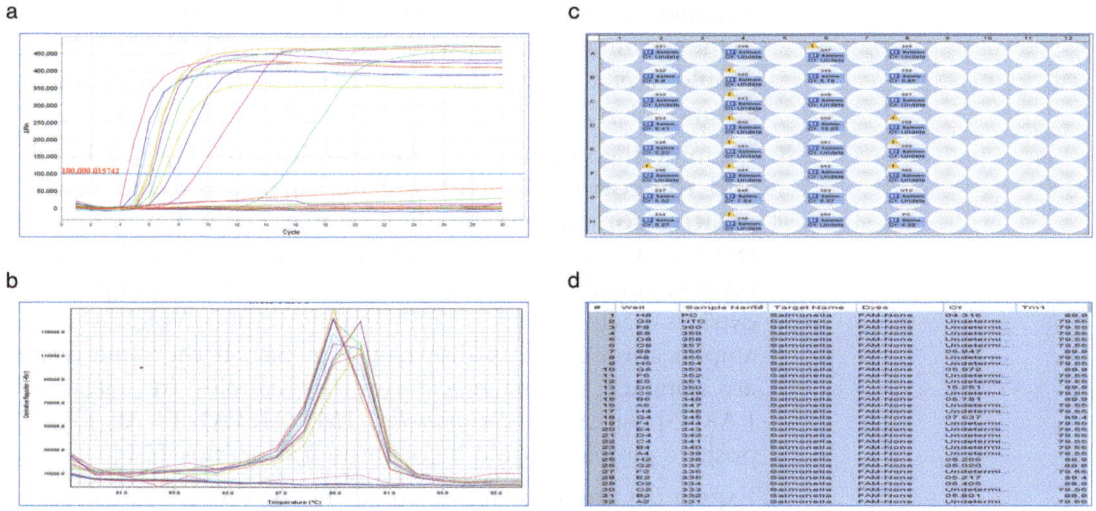

Fig. 5 Representative LAMP results viewed in the 7500 software (v2.3). In this LAMP run, A2 to H2, A4 to H4, A6 to H6, and A8 to F8 are frozen fish samples 331–360. H8 and G8 are PC and NTC, respectively. PC is *S.* Typhimurium ATCC 19585 (LT2) at ca. 10^4 cells per reaction and NTC is molecular grade water. Panel (**a**) shows the [Amplification Plot] tab for fluorescence readings during LAMP amplification. Panel (**b**) shows the [Melt Curve] tab for fluorescence readings during the melt curve stage. Panel (**c**) shows the [View Plate Layout] tab for C_T values for each well on the plate. Panel (**d**) shows the [View Well Table] tab for a tabular view of the LAMP results

3.5.3 7500 Fast: Interpret LAMP Results Using the 7500 Software (v2.3)

1. Open the 7500 software (*see* **Note 29**).
2. Open the experiment file (.eds) for the LAMP run of interest and click the [Analysis] tab on the left panel.
3. Click the [Amplification Plot] tab (Fig. 5a), and choose "Log" as the [Graph Type] to view a graphic display of ΔRn over cycle in log scale. Observe the curve to see if the auto threshold is placed within the exponential phase of the amplification curve. If not, uncheck "Auto" next to [Threshold] and adjust it to be within the exponential phase of the amplification curve according to the instrument manual. After setting the threshold, choose "Linear" as the [Graph Type] to view a graphic display of ΔRn over cycle in linear scale.
4. Click the [Melt Curve] tab (Fig. 5b), and select "Derivative Reporter" as the [Plot] type to view a graphic display of "Derivative Reporter (−Rn′)" over temperature.
5. Click the [View Plate Layout] tab (Fig. 5c), select wells to see C_T values (cycle threshold; cycle), and for each well, check corresponding [Amplification Plot] and [Melt Curve] tabs (*see* **Note 30**).
6. Click the [View Well Table] tab (Fig. 5d), which shows a tabular view of the results for each well, including C_T and T_m (melting temperature; °C) (*see* **Note 30**).

7. Examine the control wells first. The NTC well should report "Undetermined" for C_T while T_m can be <83 °C. The PC well should have a C_T less than 10 cycles (1 min per cycle) and T_m around 89 ± 2 °C.
8. Examine the sample wells. **All samples with $C_T \leq 20$ cycles and the correct T_m (89 ± 2 °C) are considered LAMP-positive for *Salmonella*.** For samples with the correct T_m (89 ± 2 °C) but $C_T > 20$ cycles or "undetermined," manually observe amplification curves and repeat LAMP on sample DNA extracts having valid LAMP amplifications (typical sigmoid amplification curve as shown in Fig. 5a). If the same results are obtained for the repeat run, the samples are considered LAMP-positive for *Salmonella*. **IMPORTANT**: All LAMP-positive samples should be confirmed by culture [21].
9. Export data to a .txt file or an .xls file or PowerPoint slides or print the results in pdf format.

4 Notes

1. Use locking tubes for the boiling step during DNA extraction to keep the caps from popping open and prevent cross-contamination.
2. Isopropanol (70%) is used as a disinfectant to clean the work surface.
3. A DNA- and DNase-degrading solution such as DNA AWAY is essential for cleaning the work surface before setting up molecular-based assays, especially for the ones that are very sensitive in the case of LAMP.
4. Other reagent choices include GspSSD Isothermal Mastermix (OptiGene Ltd.) and WarmStart LAMP Kit (New England Biolabs, Ipswich, MA).
5. If possible, use nuclease-free molecular grade water. Aliquot molecular grade water to prevent cross-contamination.
6. LAMP primers are synthesized with standard desalting purification.
7. The stock solutions of LAMP primers and LAMP primer mix (10×) may be stored at −80 °C for up to 2 years.
8. A PC and an NTC should be included in every LAMP run. Alternatively, a positive culture control may be used as the PC. Inoculate BPW at the time of sample enrichment with a *Salmonella* reference strain, e.g., *Salmonella enterica* serovar Gaminara Sal57 (a green fluorescent protein [GFP]-producing strain derived from FDA SAL5695; Microbiologics, St. Cloud,

MN), incubate the broth, and extract DNA concurrently with the animal food samples (see Subheading 3.1) consistent with laboratory QA/QC procedures.

9. Use the same molecular grade water aliquot that was used in setting up the LAMP reaction mix. This helps with troubleshooting in the event of cross-contamination.
10. Preferably, do not thaw frozen fish samples before analysis. If frozen fish sample must be tempered to obtain analytical portion, thaw below 45 °C for <15 min with continuous agitation in thermostatically controlled water bath or thaw within 18 h at 2–5 °C. Sterilize the outside of the package first, and then use sterile knives or other appropriate tools to cut samples while frozen.
11. Blend the samples at 10,000–12,000 rpm or at low-speed setting if the blending speed cannot be adjusted.
12. Wipe down the pipettes between samples with 70% isopropanol. If some of the enriched samples have *Salmonella*, moving that pipette from jar to jar can be a cross-contamination risk.
13. Use sterile scoops and either change gloves between samples or wipe down bag and gloves with 70% isopropanol between samples to prevent cross-contamination.
14. The dry dog food sample is placed at one side of the filter in the bag. After enrichment and mixing with BPW, a 1-mL liquid is taken from the other side of the filter for DNA extraction. The practice can prevent dry dog food sample particles get into the DNA extraction creating inconsistency and higher inhibition during the amplification. Also, wipe down the pipettes between samples with isopropanol. If some of the enriched samples have *Salmonella*, moving that pipette from bag to bag can be a cross-contamination risk.
15. Preferably, work in a nucleic acid workstation (dead air box) with a UV light to disinfect between assay preps.
16. Always include a PC and an NTC in every LAMP run. When using only one Genie II block (A or B, 8 samples total), prepare the LAMP reaction mix according to the column for 9 reactions (1 extra for pipetting loss). When using both Genie II blocks (A and B, 16 samples total), prepare the LAMP reaction mix according to the column for 18 reactions (2 extras for pipetting loss). For all other sample numbers, adjust the volumes accordingly.
17. To prevent cross-contamination, it is highly recommended to physically separate the areas used for preparing the LAMP reaction mix and adding DNA templates, either in different areas of the same laboratory or in different laboratories. Use a

separate set of pipettes and aerosol-resistant tips. Wipe down the pipettes with DNA AWAY.

18. The default LAMP run profile consists of amplification at 65 °C for 30 min and an anneal phase from 98 °C to 80 °C with 0.05 °C decrement per sec. Optionally, a modified LAMP run profile may be saved for later use. Tap the [Save] icon and give the profile a unique name. Tap the [Folder] icon at the bottom left of the home screen and select [Profile] to load the saved profile.
19. Wait for run timer to start counting down before you walk away. Sometimes the software will get snagged at this point.
20. The Genie II is a stand-alone instrument that can be used for the entire LAMP analysis including the run and data interpretation. It can be connected to a computer to analyze LAMP run data with computer software (Genie Explorer). LAMP run data generated on 7500 Fast can only be analyzed with computer software (7500 software).
21. Check the manufacturer's site (http://www.optigene.co.uk/support/) periodically for Genie II instrument firmware updates and communicate with the manufacturer for questions or troubleshooting.
22. The [Profile] and [Temperature] tabs show programmed and actual temperatures, respectively, in the sample wells as the LAMP reaction proceeds. The [Amplification] and [Anneal] tabs show fluorescence readings and changes in fluorescence during the amplification and anneal phases, respectively. The [Results] tab shows a tabular view of the LAMP results. The Genie II instrument takes a florescence reading every 15 s during the isothermal phase and continuous readings every second during the anneal phase.
23. The T_a values may differ slightly based on the matrix being tested, the *Salmonella* serovar, or the instrument being used.
24. As Genie II offers stand-alone operation, a computer connecting to the instrument is not required, but good for running analysis remotely and saving images from run results for presentation or publication use. Check the manufacturer's site (http://www.optigene.co.uk/support/) periodically for Genie software updates and communicate with the manufacturer for questions or troubleshooting.
25. The LAMP runs are automatically saved and organized by date in the instrument. Data can be transferred from the instrument to a computer by a USB memory stick.
26. Similar to the Genie II instrument panel view, the [Profile] and [Temperature] tabs show programmed profiles and actual

temperatures, respectively, in the sample wells as the LAMP reaction proceeds. The [Amplification/Amplification Rate] and [Anneal/Anneal Derivative] tabs show fluorescence readings or changes in fluorescence during the amplification and anneal phases, respectively. The [Results] tab shows a tabular view of the LAMP results that differ slightly from the instrument panel view.

27. The adjustment is necessary to ensure that all valid peaks are identified, and the results obtained using the software match with those displayed on the instrument panel.
28. Similar to the Genie II instrument panel view, there are four columns (Graph Name, Well Number, Well Name, and Peak Value). The top portion of the "Peak Value" column shows "Amp Time" for each sample ("Well Name"), while the bottom portion shows "Anneal Derivative" for any amplified product in that well.
29. The LAMP run data generated on 7500 Fast can be analyzed using the 7500 software, which can be used as a stand-alone software for remote analysis.
30. Samples with undetermined C_T may be flagged in [View Plate Layout] and [View Well Table], which is common and acceptable. It is a good practice to double check the corresponding [Amplification Plot] and [Melt Curve] tabs for these samples.

Acknowledgments

The authors thank Paul Park, Cindy Wu, Tameji Eames, and David Kiang from California Department of Public Health, Food and Drug Laboratory Branch for providing LAMP run graphs on Genie II when analyzing dry dog food samples in a multi-laboratory validation study. The authors are grateful to members of the FDA's Microbiology Methods Validation Subcommittee (MMVS) and Bacteriological Analytical Manual (BAM) Council for critically reviewing the *Salmonella* LAMP method validation studies.

Disclaimer The views expressed in this chapter are those of the authors and may not reflect the official policy of the FDA, the Department of Health and Human Services, or the US Government. Reference to any commercial materials, equipment, or process does not in any way constitute approval, endorsement, or recommendation by the FDA.

References

1. Notomi T, Okayama H, Masubuchi H, Yonekawa T, Watanabe K, Amino N et al (2000) Loop-mediated isothermal amplification of DNA. Nucleic Acids Res 28(12):E63. https://doi.org/10.1093/nar/28.12.e63
2. Soroka M, Wasowicz B, Rymaszewska A (2021) Loop-mediated isothermal amplification (LAMP): the better sibling of PCR? Cells 10(8). https://doi.org/10.3390/cells10081931
3. Mori Y, Notomi T (2020) Loop-mediated isothermal amplification (LAMP): expansion of its practical application as a tool to achieve universal health coverage. J Infect Chemother 26(1): 13–17. https://doi.org/10.1016/j.jiac.2019.07.020
4. Kumar Y, Bansal S, Jaiswal P (2017) Loop-mediated isothermal amplification (LAMP): a rapid and sensitive tool for quality assessment of meat products. Compr Rev Food Sci F 16(6):1359–1378. https://doi.org/10.1111/1541-4337.12309
5. Mansour SM, Ali H, Chase CC, Cepica A (2015) Loop-mediated isothermal amplification for diagnosis of 18 world Organization for Animal Health (OIE) notifiable viral diseases of ruminants, swine and poultry. Anim Health Res Rev 16(2):89–106. https://doi.org/10.1017/S1466252315000018
6. Kaneko H, Kawana T, Fukushima E, Suzutani T (2007) Tolerance of Loop-mediated isothermal amplification to a culture medium and biological substances. J Biochem Biophys Methods 70(3):499–501
7. Francois P, Tangomo M, Hibbs J, Bonetti EJ, Boehme CC, Notomi T et al (2011) Robustness of a Loop-mediated isothermal amplification reaction for diagnostic applications. FEMS Immunol Med Microbiol 62(1):41–48. https://doi.org/10.1111/j.1574-695X.2011.00785.x
8. Yang Q, Wang F, Prinyawiwatkul W, Ge B (2014) Robustness of *Salmonella* loop-mediated isothermal amplification assays for food applications. J Appl Microbiol 116(1): 81–88. https://doi.org/10.1111/jam.12340
9. Yang Q, Domesle KJ, Ge B (2018) Loop-mediated isothermal amplification for *Salmonella* detection in food and feed: current applications and future directions. Foodborne Pathog Dis 15(6):309–331. https://doi.org/10.1089/fpd.2018.2445
10. Galan JE, Ginocchio C, Costeas P (1992) Molecular and functional characterization of the *Salmonella* invasion gene *invA*: homology of InvA to members of a new protein family. J Bacteriol 174(13):4338–4349. https://doi.org/10.1128/jb.174.13.4338-4349.1992
11. Chen S, Wang F, Beaulieu JC, Stein RE, Ge B (2011) Rapid detection of viable *Salmonellae* in produce by coupling propidium monoazide with Loop-mediated isothermal amplification. Appl Environ Microbiol 77(12):4008–4016. https://doi.org/10.1128/AEM.00354-11
12. Yang Q, Chen S, Ge B (2013) Detecting *Salmonella* serovars in shell eggs by Loop-mediated isothermal amplification. J Food Prot 76(10):1790–1796. https://doi.org/10.4315/0362-028X.JFP-13-140
13. Yang Q, Wang F, Jones KL, Meng J, Prinyawiwatkul W, Ge B (2015) Evaluation of Loop-mediated isothermal amplification for the rapid, reliable, and robust detection of *Salmonella* in produce. Food Microbiol 46:485–493. https://doi.org/10.1016/j.fm.2014.09.011
14. Yang Q, Domesle KJ, Wang F, Ge B (2016) Rapid detection of *Salmonella* in food and feed by coupling Loop-mediated isothermal amplification with bioluminescent assay in real-time. BMC Microbiol 16(1):112. https://doi.org/10.1186/s12866-016-0730-7
15. Domesle KJ, Yang Q, Hammack TS, Ge B (2018) Validation of a *Salmonella* Loop-mediated isothermal amplification assay in animal food. Int J Food Microbiol 264:63–76. https://doi.org/10.1016/j.ijfoodmicro.2017.10.020
16. Domesle KJ, Young SR, Ge B (2021) Rapid screening for *Salmonella* in raw pet food by Loop-mediated isothermal amplification. J Food Prot 84(3):399–407. https://doi.org/10.4315/JFP-20-365
17. ISO (2016) ISO 16140-2:2016 microbiology of the food chain – method validation – part 2: protocol for the validation of alternative (proprietary) methods against a reference method. https://www.iso.org/standard/54870.html. Accessed 22 July 2024
18. Domesle KJ, Young SR, McDonald RC, Ge B (2022) Versatility of a *Salmonella* Loop-mediated isothermal amplification assay using multiple platforms and master mixes in animal food matrices. J AOAC Int 105(6):

1503–1515. https://doi.org/10.1093/jaoacint/qsac059

19. Ge B, Domesle KJ, Young SR, McDonald RC, Pamboukian RY (2025) Independent laboratory validation of a *Salmonella* Loop-mediated isothermal amplification assay in 27 human and animal food matrices of 9 ISO food categories. Food Microbiol in press
20. Ge B, Domesle KJ, Yang Q, Hammack TS, Wang SS, Deng X et al (2019) Multilaboratory validation of a Loop-mediated isothermal amplification method for screening *Salmonella* in animal food. Front Microbiol 10:562. https://doi.org/10.3389/fmicb.2019.00562
21. Young SR, Domesle KJ, McDonald RC, Lozinak KA, Laksanalamai P, Harrell E et al (2022) Toward the adoption of Loop-mediated isothermal amplification for *Salmonella* screening at the national antimicrobial resistance monitoring system's retail meat sites. Foodborne Pathog Dis 19(11):758–766. https://doi.org/10.1089/fpd.2022.0047
22. Andrews WH, Wang H, Jacobson A, Ge B, Zhang G, Hammack TS (2024) Bacteriological Analytical Manual Chapter 5: *Salmonella*. https://www.fda.gov/food/laboratory-methods-food/bam-chapter-5-salmonella. Accessed 22 July 2024

Chapter 3

Design of a Competitive Internal Amplification Control in Multiplex Loop-Mediated Isothermal Amplification

Foteini Roumani, Alexandre Lamas, and Alejandro Garrido-Maestu

Abstract

Loop-mediated isothermal amplification, LAMP, has attracted the interest of the scientific community due to several of its advantages over classical nucleic acid amplification (NAA) techniques, namely, PCR and qPCR, as well as over other isothermal techniques such as NASBA, HDA, and/or RPA. However, the design of multiplex assays remains challenging. This issue is critical if considered that for specific applications, an amplification control, which ideally would be internal (IAC), is already mandatory for methods based on the application of PCR/qPCR. In the current chapter, we will guide the reader on how to design a competitive IAC for its successful implementation in a multiplex assay.

Key words Competitive internal amplification control, *Listeria monocytogenes*, Loop-mediated isothermal amplification, Multiplex LAMP

1 Introduction

Since its original description, back in 2000, loop-mediated isothermal amplification (LAMP) has attracted the interest of the scientific community due to its inherent advantages over other nucleic acid amplification techniques, including other isothermal [1–3]. Additionally, many scientists have exploited the advantages of LAMP for the design and fabrication of different types of miniaturized and point-of-care (POC) devices [4–6] to be applied in medical setups, food processing facilities, and decentralized laboratories, among others [7]. This widespread adoption has already generated the appearance of commercial products based on this technology, as well as the release of scientific publications covering interlaboratory validation of open formula assays [8, 9]. One key aspect for the wider acceptance of LAMP-based methods for food analyses relies on the implementation of proper controls to assure negative results are truly negative, and not the result of reaction inhibition. Even though not many assays have implemented such controls, the ones

Alejandro Garrido-Maestu and Alexandre Lamas (eds.), *Foodborne Pathogens: Loop-mediated Isothermal Amplification Protocols*, Methods and Protocols in Food Science, https://doi.org/10.1007/978-1-0716-4526-0_3,

most reported rely on the amplification of another gene/target in a parallel reaction [10, 11]. However this approach increases the cost per sample as well as the hands-on work, the likelihood of errors, and of outmost importance, it does not fully assure lack of reaction inhibition in a given reaction as it is performed in different tube [12]. A rational way to proceed would be to implement an internal amplification control (IAC), which is nowadays already included in most, if not all, PCR/qPCR assays [13–15]. Unfortunately, due to the complexity of primer design and relative difficulties for multiplexing assays, such controls have not been widely implemented in LAMP assays and very limited number of assays have reported them [8, 16, 17]. In the present chapter, the reader will be guided through the steps needed to design a competitive IAC suitable for easy implementation in a multiplex LAMP assay. As an example, throughout the chapter, *L. monocytogenes* will be used as model organism for the design; however it is noteworthy that the same procedure was reported successful for other foodborne and clinical pathogens [8, 16, 17].

2 Materials

2.1 Bacteria and Media

1. *E. coli* JM109 competent cells (Promega, Madison, WI, USA).
2. Luria Bertani (PanReac AppliChem, Barcelona, Spain).

2.2 Chemicals

Unless otherwise stated, all the chemicals were purchased from Sigma–Aldrich (St. Louis, USA).

1. Phosphate buffer saline with Tween® 20 (PBST): 155.7 mM NaCl, 2.8 mM KCl, 2.97 mM Na_2HPO_4, 1.06 mM KH_2PO_4, 0.5 g Tween® 20, 1000 mL Milli-Q water. Final pH $= 7.4 \pm 0.2$.
2. Tris-EDTA solution (TE 1×): 10 mM Tris-HCl, 1 mM EDTA. Final pH $= 8.0 \pm 0.2$.
3. Enzymatic solution: 1 mg/mL lysozyme 1 mg/mL achromopeptidase, dissolved in TE 1×.
4. NucleoSpin Plasmid kit (Macherey-Nagel, Düren, Germany).
5. Primers (*see* Table 1).
6. GspSSD Isothermal Master Mix ISO-004 (OptiGene Ltd., Horsham, UK).
7. ROX, passive reference dye (Invitrogen, Carlsbad, CA, USA).

2.3 Equipment

1. StepOne Plus™ real-time PCR system (Applied Biosystems™, Foster City, CA, USA).
2. Centrifuge with a speed range of 380–16,000 × *g* and a rotor suitable for 1.5–2 mL microtubes.

Table 1
Primers

Primer name	Sequence 5′ → 3′	Concentration	References
plcA-FIP	GCA GCG CTC TCT ATA CCA GGT ACA *tttt* AAT GTC CAT GTT ATG TCT CCG TTA	1000	[24, 25]
plcA-BIP	AGG TTT GTT GTG TCA GGT AGA GCG *tttt* CG C TTA ATA ACT GGA ATA AGC CAA	1000	
plcA-F3	TGT GTT TGA GCT AGT GGT TTG G	200	
plcA-B3	CCC ATT AGG CGG AAA AGC ATA T	200	
plcA-LB	CAT CCA TTG TTT TGT AGT TAC AGA G	600	

The "-*tttt*-" sequence in FIP and BIP serves as a linker among the F2-F1c and B2-B1c primers, respectively

3. Heating block (dry bath) capable of shaking (1400 rpm) and heating (99 °C). In the described methodology, a Thermomixer 5436 (Eppendorf AG, Germany) was used.
4. Qubit™4 Fluorometer (Invitrogen™, Carlsbad, CA, USA 2.4).

2.4 Software

1. Primer Explorer (https://primerexplorer.jp/e/index.html).
2. Geneious Prime® version 2023.0.1 (Biomatters Ltd., Auckland, New Zealand. https://www.geneious.com).
3. Oligo Calc: Oligonucleotide Properties Calculator (http://biotools.nubic.northwestern.edu/OligoCalc.html) [18].
4. Random DNA generator [19].
5. uMELT [20].

3 Methods

The IAC developed by Roumani et al. included in their assay for the detection of *L. monocytogenes* was selected as a model to guide the reader through the process [17].

3.1 Sequence Design

1. Select a target gene of interest and retrieve its sequence from a reference database in fasta format (*see* **Note 1**). In this case, the gene *plcA* was selected for the detection of *L. monocytogenes.*
2. Copy the sequence in a text processing software, Word, Notepad, or similar, and save it as a .txt file (*see* **Note 2**).
3. Upload the .txt file to Primer Explorer (https://primerexplorer.jp/lampv5e/index.html) or any other suitable software for LAMP primer design (*see* **Note 3**), and click on "Primer design." Select "Generate" for a basic design that tends to provide suitable primers (advanced LAMP users may

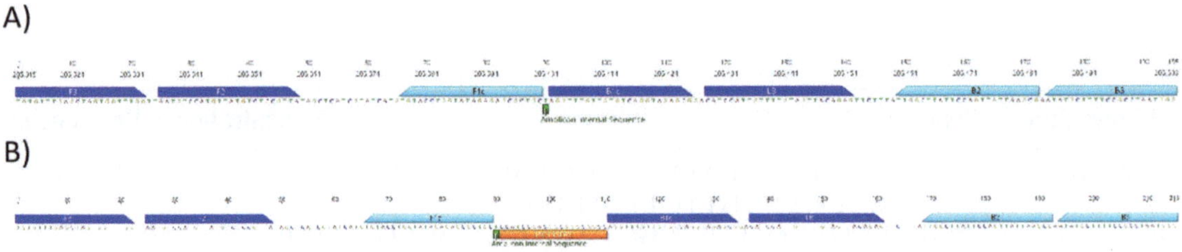

Fig. 1 (**a**) Original *plcA* gene sequence with internal region highlighted and LAMP primers positioned. (**b**) *plcA* sequence with the chimeric IAC fragment highlighted in orange

perform this step with the "Detail Settings" option). Then click on "Display" to see the candidate primer sequences. Once a suitable set of FIP, BIP, F3, and B3 primers is selected, download the corresponding files and use the "Primer Information" file for the design of loop primers; *see* **Notes 4** and **5**. The procedure for this second part is the same as detailed for the other primers, but instead of uploading the .txt file, now the "Primer Information" file is selected, and after hitting on "Primer Design" in the same screen with the sequence of the gene, the positions of FIP, BIP, F3, and B3 will be indicated; now the same procedure can be followed for the design of loop primers.

4. Import the sequences of each primer in a suitable analysis software such as GENEious, CLC Sequence viewer, or similar, and identify their position in the reference sequence (the sequence used to generate the .txt file selected for primer design). Select the fragment between F3 and B3, copy it, and paste it in http://biotools.nubic.northwestern.edu/OligoCalc.html#helpIUPAC; by selecting "Calculate" the length, %GC and melting temperature (T_m) of the sequence will be provided. This will be the Tm of the target, *plcA* in the present example.

5. The same primers previously designed will be used for the IAC; thus no additional design is needed. Going back to the sequence analysis software, now we must focus on the sequence between F1c and B1c, in this case a single "T"; see Fig. 1. Now a random sequence may be generated by hand or a specific program such as FaBox (https://users-birc.au.dk/palle/php/fabox/random_sequence_generator.php) may be used [19]. This new sequence will be rich in GC so that with a small fragment, a significant Tm increase may be achieved. A piece of this random sequence can be introduced in the software between F1c and B1c and the new size, %GC, and Tm will be recalculated following the procedure described above. In this example, the sequence selected was "CCA TCC GAC GGC GGG CGG CC" as depicted in Fig. 1b. Thus, the

fragment increased from 195 bp to 215 bp and the %GC from 42% to 46%, resulting in a Tm increase from a theoretical 78.5–80.2 °C.

6. Assess if the Tm difference is enough with uMELT (https://www.dna-utah.org/umelt/quartz/umb.php#close); see Fig. 2a, b [20].
7. The sequence with the modified inner region can now be purchased from any commercial supplier. The preferred format would be cloned into a plasmid as it could be used to transform competent bacteria for long-term storage and will also allow for a more accurate quantification.
8. With the plasmid, e.g., pUC57, competent bacteria such as *E. coli* JM109 can be transformed following the standard protocol provided by the supplier.
9. Pure, transformed bacterial cultures can be used now for the recovery of the plasmid that will serve as DNA for the IAC. The extraction can be performed with a plasmid-specific extraction kit such as NucleoSpin Plasmid following the manufacturer's instructions.
10. Quantify the concentration of plasmidic DNA with a Qubit and, if desired, convert it in copies. Given the fact that the size is known and the amount of ng was just determined, the conversion can be simply done with existing online tools such as https://www.scienceprimer.com/copy-number-calculator-for-realtime-pcr.
11. The plasmid can now be used as IAC in our LAMP assay. Note that it has the same primers as our target, *plcA*, but the size of the amplicon is larger, and attending to the data obtained must have a higher Tm that we should be able to differentiate after melt curve analysis.
12. Prepare a LAMP reaction as detailed in Table 2. Set a real-time PCR thermocycler, or other suitable machine (*see* **Note 6**), to detect in the FAM/SYBR channel with ROX as passive reference dye (*see* **Note 7**), and configure one single step consisting of 60 cycles of 30 s with signal acquisition after each one, with a final melt curve analysis step, in Table 3 details for temperature setup are provided (*see* **Note 8**).

3.2 Result Interpretation

1. Confirm an amplification plot is visible. If this is not the case, it will mean that reaction inhibition has occurred.
2. If amplification plot appears, then move to the melt curve analysis tab to evaluate the Tm of the amplicon.
3. Attending to the bioinformatic analysis, in this example, one would expect for a *L. monocytogenes* positive sample a peak at 78.5 °C to and at 80.2 °C for the IAC. This temperature

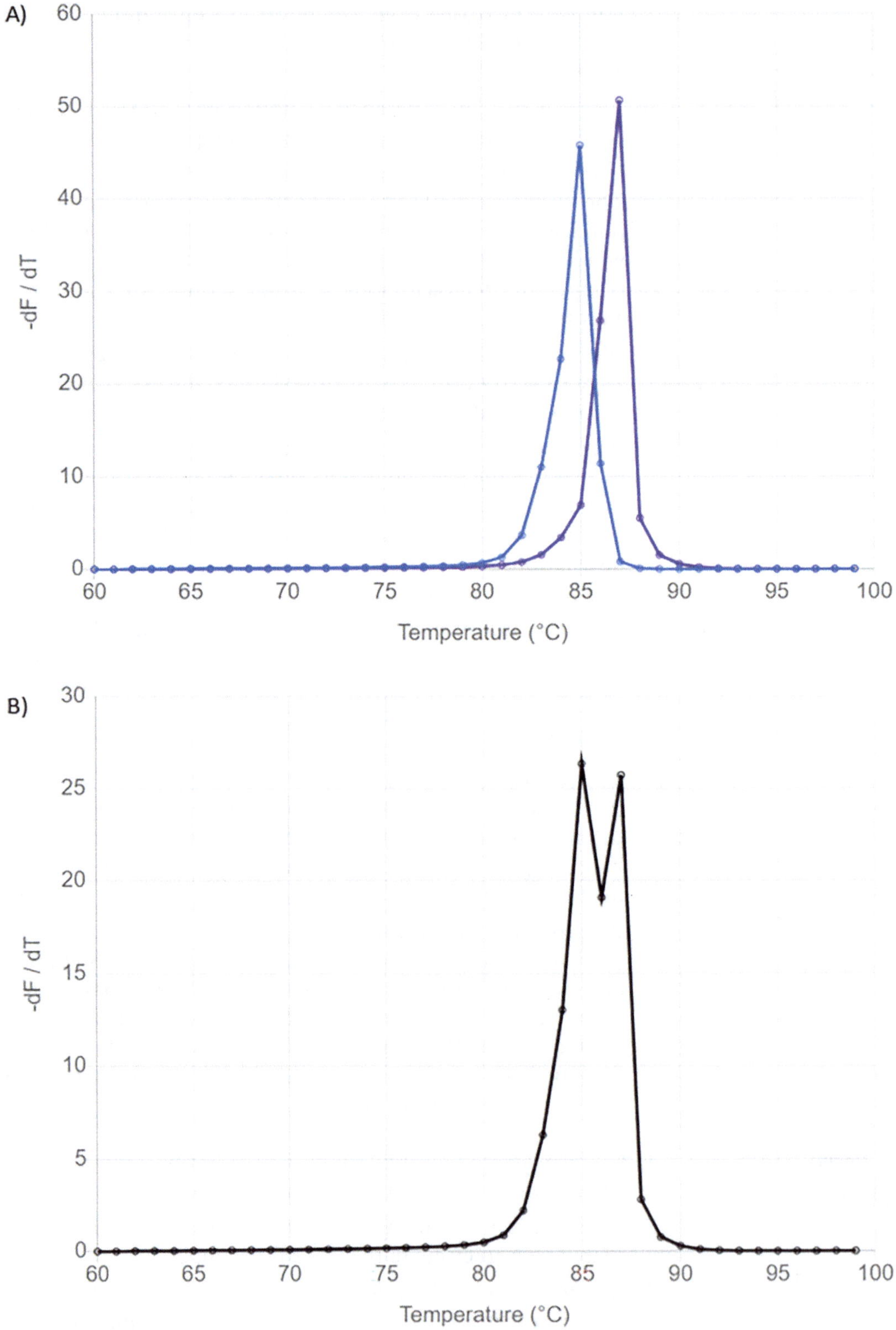

Fig. 2 (**a**) uMELT predicted melt curves for the original target, *plcA* amplicon in blue, and the after the insertion of the chimeric DNA in purple. (**b**) Combined melt curve of the target and the IAC amplicons

Table 2
LAMP reaction mix preparation

Primer/probe	Volume/rxn (μL)	Final concentration (nM)
Master mix	15.0	–
plcA-FIP	0.25	1000
plcA-BIP	0.25	1000
plcA-F3	0.05	200
plcA-B3	0.05	200
plcA-LB	0.15	600
IAC DNA	1.0	760[a]
ROX	0.05	50
H_2O	2.20	–
Template DNA	6.0	–
Final volume	25	–

Primer volumes indicated are calculated based on a working solution of 100 μM; *see* **Note 10**

[a]Copies, the volume may vary depending on DNA stock concentration

Table 3
Thermal profile for multiplex qPCR

Step	Temperature (°C)	Time (s)	Cycles
Amplification	65[a]	30	60
Melt curve	95 80[b] 95	15 60 1	1

[a]Detect fluorescence after this step in every cycle

[b]The temperature increments for fluorescence acquisition can be set at 0.2 °C/ s; *see* **Note 11**

difference should be easily differentiated by melt curve analysis attending to the uMELT analysis; however, this value has to be experimentally determined.

4. The actual Tm values recovered by Roumani et al. for *L. monocytogenes* and the IAC were 84.8 ± 0.6 °C and 87.6 ± 0.8 °C.
5. Attending to the experimental values calculated, a positive sample will present one peak at ~85 °C and may or may not have a second peak at ~88 °C (*see* **Note 9**). A negative sample will only have one single peak at ~88 °C corresponding to the IAC. Table 4 provides a summary of the assays result interpretation attending to the target-IAC combination.

Table 4
LAMP result interpretation

Target gene		Internal amplification control (IAC) Positive	Negative
plcA	Positive	Presence of *L. monocytogenes*/25 g	Presence of *L. monocytogenes*/25 g[a]
	Negative	Absence of *L. monocytogenes*/25 g	LAMP Inhibition[b]

[a]Samples with high concentration of target DNA may result in no amplification of the IAC
[b]*See* **Note 12**

4 Notes

1. Whenever possible retrieve sequence data from high-quality, curated databases such as RefSeq from NCBI (https://www.ncbi.nlm.nih.gov/refseq/).
2. The latest version of Primer Explorer only allows to upload sequences 1000 bp long. If the region of interest is longer, it will have to be split in different files and primer design will be performed independently for each file.
3. Primer Explorer was the first specific software, and one of the most common ones, for LAMP primer design. However, nowadays other alternatives are available such as LAMP Designer (PREMIER Biosoft, https://www.premierbiosoft.com/isothermal/lamp.html), MorphoCatcher [21], LAVA [22], or NEB® LAMP Primer Design Tool (https://lamp.neb.com/#!/).
4. Loop primers are not essential for the successful performance of LAMP assays. However, they significantly improve the reaction speed and it is a good practice to implement both, forward and reverse, whenever possible [23].
5. It is recommended to purchase all primers HPLC purified. However, if there are budget constraints, F3, B3, LF, and LB may be acquired with a standard desalting purification, and only request the HPLC purification for FIP and BIP.
6. A fluorimeter and other portable commercial devices are also suitable such as the Genie® II, Genie® III, Genie® HT (https://www.optigene.co.uk/instruments/), Axxin T8-ISO and Axxin T16-ISO (https://www.axxin.com/Molecular-T8.php, https://www.axxin.com/Molecular-T16.php), and the BioRanger (https://shop.diagenetix.com/product-p/sd.v.3.0.htm), among others.
7. ROX is a passive reference dye used to provide a more stable baseline; however, the amplification reaction may be performed without it or with a different dye upon proper configuration in the corresponding software.

8. The overall amplification time must be 30 min, for this different number of cycles and amplification time per cycle can be set. Even though this may change the shape of the final amplification plot as it may have more, or less, data points, the result interpretation will not change. In a similar way, modifying the starting temperature, and the increments, in the melt curve analysis step, may modify the shape of the peaks, but the interpretation will be the same.
9. The IAC peak may not visible in samples with a high concentration of the target.
10. A 10× primer stock may be prepared so that, instead of adding each primer individually, 2.5 μL of this stock can be used. This will cause a readjustment in the volume of nuclease-free water to be added per reaction to 0.45 μL (Table 2).
11. Modifying temperature increments, or setting fluorescence acquisition in "continuous mode," may generate minor differences in the melting analysis. The Tm value should stay very similar however the plot may look different (fluorescence acquisition modes in melt curve analysis may vary from instrument to instrument).
12. Reaction inhibition may typically be overcome by diluting the sample in sterile, nuclease-free water, TE 1×, or any other suitable diluent. Depending on the degree of inhibition, and the inhibitory compound, diluting 1/2 or 1/10 tends to be enough to recover amplification of the IAC and potentially of the target if present in the sample.

References

1. Garrido-Maestu A, Prado M (2022) Naked-eye detection strategies coupled with isothermal nucleic acid amplification techniques for the detection of human pathogens. Compr Rev Food Sci Food Saf
2. Li Y, Fan P, Zhou S et al (2017) Loop-mediated isothermal amplification (LAMP): a novel rapid detection platform for pathogens. Microb Pathog 107:54–61
3. Garg N, Ahmad FJ, Kar S (2022) Recent advances in Loop-mediated isothermal amplification (LAMP) for rapid and efficient detection of pathogens. Curr Res Microb Sci 3:100120
4. Njiru ZK (2012) Loop-mediated isothermal amplification technology: towards point of care diagnostics. PLoS Negl Trop Dis 6:1–4
5. Zhang H, Xu Y, Fohlerova Z et al (2019) LAMP-on-a-chip: revising microfluidic platforms for Loop-mediated DNA amplification. Trends Anal Chem 113:44–53
6. Kaygusuz D, Vural S, Aytekin AÖ et al (2019) DaimonDNA: a portable, low-cost Loop-mediated isothermal amplification platform for naked-eye detection of genetically modified organisms in resource-limited settings. Biosens Bioelectron 141
7. Hassan MM, Grist LF, Poirier AC et al (2022) Loop-mediated isothermal amplification (LAMP): for the rapid detection of nucleic acid targets in resource-limited settings. J Med Microbiol:1–6
8. D'Agostino M, Diez-Valcarce M, Robles S et al (2015) A Loop-mediated isothermal amplification-based method for analysing animal feed for the presence of *Salmonella*. Food Anal Methods 8:2409–2416
9. D'Agostino M, Robles S, Hansen F et al (2016) Validation of a Loop-mediated amplification/ISO 6579-based method for analysing soya meal for the presence of *Salmonella enterica*. Food Anal Methods 9:2979–2985

10. Roumani F, Azinheiro S, Sousa H et al (2021) Optimization and clinical evaluation of a multi-target Loop-mediated isothermal amplification assay for the detection of SARS-CoV-2 in nasopharyngeal samples. Viruses 13
11. Mautner L, Baillie CK, Herold HM et al (2020) Rapid point-of-care detection of SARS-CoV-2 using reverse transcription Loop-mediated isothermal amplification (RT-LAMP). Virol J 17:1–14
12. Sritong N, Ngo WW, Ejendal KFK et al (2023) Development of an integrated sample amplification control for salivary point-of-care pathogen testing. Anal Chim Acta 1287:1–23
13. Hoorfar J, Cook N, Malorny B et al (2003) Making internal amplification control mandatory for diagnostic PCR. J Clin Microbiol 41: 5835
14. Hoorfar J, Cook N, Malorny B et al (2004) Diagnostic PCR: making internal amplification control mandatory. Lett Appl Microbiol 96: 221–222
15. Deer DM, Lampel KA, González-Escalona N (2010) A versatile internal control for use as DNA in real-time PCR and as RNA in real-time reverse transcription PCR assays. Lett Appl Microbiol 50:366–372
16. Bath C, Scott M, Sharma PM et al (2020) Further development of a reverse-transcription Loop-mediated isothermal amplification (RT-LAMP) assay for the detection of foot-and-mouth disease virus and validation in the field with use of an internal positive control. Transbound Emerg Dis 67:2494–2506
17. Roumani F, Azinheiro S, Carvalho J et al (2021) Loop-mediated isothermal amplification combined with immunomagnetic separation and propidium monoazide for the specific detection of viable *Listeria monocytogenes* in milk products, with an internal amplification control. Food Control 125:107975
18. Kibbe WA (2007) OligoCalc: An online oligonucleotide properties calculator. Nucleic Acids Res 35:43–46
19. Villesen P (2007) FaBox: an online toolbox for FASTA sequences. Mol Ecol Notes 7:965–968
20. Dwight Z, Palais R, Wittwer CT (2011) uMELT: prediction of high-resolution melting curves and dynamic melting profiles of PCR products in a rich web application. Bioinformatics 27:1019–1020
21. Shirshikov FV, Pekov YA, Miroshnikov KA (2019) MorphoCatcher : a multiple-alignment based web tool for target selection and designing taxon-specific primers in the loop-mediated isothermal amplification method. PeerJ 7: e6801
22. Torres C, Vitalis EA, Baker BR et al (2011) LAVA: an open-source approach to designing LAMP (Loop-mediated isothermal amplification) DNA signatures. BMC Bioinformatics 12:240
23. Nagamine K, Hase T, Notomi T (2002) Accelerated reaction by Loop-mediated isothermal amplification using loop primers. Mol Cell Probes 16:223–229
24. Garrido-Maestu A, Azinheiro S, Carvalho J et al (2018) Development and evaluation of Loop-mediated isothermal amplification, and recombinase polymerase amplification methodologies, for the detection of *Listeria monocytogenes* in ready-to-eat food samples. Food Control 86
25. Azinheiro S, Roumani F, Prado M et al (2022) Rapid same-day detection of *Listeria monocytogenes*, *Salmonella* spp., and *Escherichia coli* O157 by colorimetric LAMP in dairy products. Food Anal Methods 15:2959–2971

Chapter 4

Detection of Thermophilic *Campylobacter* spp. in Poultry Meat

Arnaboldi Sara and Filipello Virginia

Abstract

Campylobacter spp. are gram-negative, non-spore-forming bacteria responsible for the majority of foodborne human gastroenteritis in European countries. The main reservoirs for foodborne transmission are avian species, particularly poultry, with handling and eating undercooked meat representing the main risk factor for human infection. The rapid detection and identification of *Campylobacter* in chicken meat is important to control the spread of foodborne infections. A valid alternative to traditional real-time PCR is loop-mediated isothermal amplification (LAMP) assay, a more rapid and easy tool for the sensitive and efficient detection of *Campylobacter* spp. in contaminated poultry meat. Indeed, with LAMP, DNA amplification occurs in a shorter time (less than 60 min), and it is less susceptible to inhibitors than PCR. Moreover, it is possible to evaluate the DNA amplification by the naked eye, without additional instrumentation, through the use of alternative detection systems (i.e., colorimetric reaction). Therefore, here we present a handy 1-day workflow for the timely and precise *Campylobacter* detection in chicken meat. In particular, compared to traditional protocols, we shortened the enrichment time to 6 h and we performed a colorimetric LAMP reaction at 60 °C for 45 min, able to detect down to 10^1 CFU/g of *Campylobacter* in chicken meat.

Key words *Campylobacter* spp., Poultry, Colorimetric method, Foodborne disease

1 Introduction

In recent years, enteritis and gastroenteritis cases caused by *Campylobacter* spp. have exceeded those caused by classic enteric bacteria, making *Campylobacter* the main cause of foodborne diseases in EU countries [1]. *Campylobacter* spp. are gram-negative, mobile, non-spore-forming bacteria belonging to the Campylobacteraceae family that includes 22 species, with *Campylobacter jejuni* (*C. jejuni*) and *Campylobacter coli* (*C. coli*) being the main cause of human gastroenteritis. *Campylobacter* spp. are frequently defined as thermophilic microorganisms, due to their ability to grow at pH between 6.5 and 7.5 and at temperatures between 37 °C and 42 °C [2].

Alejandro Garrido-Maestu and Alexandre Lamas (eds.), *Foodborne Pathogens: Loop-mediated Isothermal Amplification Protocols*, Methods and Protocols in Food Science, https://doi.org/10.1007/978-1-0716-4526-0_4,

Campylobacter spp. are at risk of zoonotic transmission because they can colonize a wide range of hosts (including domestic and wild animals) and can be found in many foods of animal origin [3]. In particular, the most important sources of *C. coli* and *C. jejuni* are avian species, particularly poultry; indeed, the handling of raw chicken meat, especially in households, or the consumption of undercooked meat is the main risk factor for human infection [4, 5]. Recent studies reported that chicken meat in retails is contaminated with *C. jejuni* in 60–80% of cases in Europe; thus the rapid detection of *Campylobacter* in chicken meat is essential to limit the spread of the pathogen and to avoid human infection [6, 7]. The preferred method to screen food for the presence of *Campylobacter* is through real-time PCR, amplifying specific DNA fragments, after 24 h of sample enrichment in a proper medium. Therefore, the analysis is completed after almost 48 h (from sample arrival in laboratory to the screening result), but it can last up to 6 days considering also the final cultural confirmation. This timing may be too long to match the requirements of food safety, and food business operators, highlighting the need of new sensitive, affordable, and rapid methods. Recently, other amplification methods have been developed, with loop-mediated isothermal amplification (LAMP) being one of the most commonly used isothermal method to detect bacteria, viruses, and parasites [8–14]. LAMP technique is an isothermal reaction that requires 2/3 pairs of primers to amplify the target region using a hairpin structure with stem loops at each end. LAMP performs the amplification at a steady temperature due to the use of a polymerase with high strand-displacement activity (*Bst* polymerase from *Bacillus stearothermophilus*) [9]. LAMP is highly efficient and specific due to the use of a set of primers that includes a forward inner primer (FIP), a backward inner primer (BIP), two outer primers (forward—F3 and backward—B3), corresponding to target gene sequence of six different regions, and two loop primers designed to accelerate the amplification reaction (LF and LB) [8, 15, 16]. The amplification usually occurs in a short time (less than 60 min), with up to 10^9 DNA copies generated, making it a time-effective assay useful to effectively monitor critical points of contamination along the food chain. Moreover, LAMP methodology is less susceptible to amplification inhibitors, with a great potential in the foodborne diseases field that frequently concerns complex matrices at risk of inhibition effects [16].

The amplification occurs using a common heated plate, without the need of thermocyclers normally used in PCR, making it attractive also for a possible use "on- field" and in situations where specific equipment is lacking [17]. In fact, LAMP is a low-cost and easy-operation technique, given the possibility to evaluate the DNA-target detection by the naked eye, through the use of alternative detection systems (i.e., colorimetric reaction), and without further processing with expensive laboratory equipment.

Finally, the LAMP assay was found to be 10–100-fold more sensitive than PCR with a detection limit of 10 copy, making LAMP a rapid, cost-effective, and accessible alternative method for food screening of *Campylobacter* [16, 17]. Therefore, we developed a handy 1-day workflow for the timely and precise *Campylobacter* detection in contaminated poultry meat, improving the available protocols to detect *Campylobacter* spp. [18]. In particular, we shortened the enrichment time to 6 h and we performed a colorimetric LAMP on DNA extracted from the enrichment broth, to detect *Campylobacter* in chicken meat just observing the color change of the test tube. Performing a LAMP reaction at 60 °C for 45 min permits to detect down to 10^1 CFU/g of *Campylobacter* in chicken meat after only 6 h of enrichment, instead of 24 h (needed to perform the detection with real-time PCR protocol). The detectable pathogen concentration with colorimetric LAMP meets the limits set by the EC Regulation No. 2073/2005 regarding *Campylobacter* detection in broiler carcasses (10^3 CFU/g) [19].

Colorimetric LAMP is a rapid and easy tool for the sensitive and efficient screening of chicken meat, with a promising application in the poultry industry, where the impact of fast results is significant [17].

2 Materials

2.1 Sample Preparation

1. Preston broth (selective liquid medium for *Campylobacter* spp. detection, according to ISO 10272); it is prepared with the formula:
   - Enzymatic digest of animal tissues 10.0 g/L.
   - Peptone 10.0 g/L.
   - Sodium chloride 5.0 g/L.
   - Lysed horse blood 50 mL.
   - Polymyxin B sulfate 5000 IU.
   - Rifampicin 0.01 g/L.
   - Trimethoprim lactate salt 0.01 g/L.
   - Amphotericin B 0.01 g/L.

   The final pH should be 7.4 ± 0.2 at 25 °C.

2.2 DNA Extraction

1. Chelex solution (6%) prepared with 6.0 g of Chelex 100 in 100 mL of nuclease-free water.
2. Positive extraction control: *C. jejuni* dried strain (ATCC 33291).

Table 1
Primer sequences used for the colorimetric LAMP assay

Primer name	Sequence
FIP	3′-GGACCGTGTCTCAGTTCCAGTGTGACGGATGAGACTATATAGTATCAGC TAG-5′
BIP	3′-CGGGAGGCAGCAGTAGGGAATATTGCTAAGAAAAGGAGTTTACGCTCCG-5′
F3	3′-CTGCTTAACACAAGTTGAGTAGG-5′
B3	3′-TTCCTTAGGTACCGTCAGAA-5′
LF	3’-GTTAAGCGTCATAGCCTTGGTAA-5′
LB	3′-GCGTGGAGGATGACACTT-5’

2.3 LAMP Reaction

1. WarmStart Colorimetric LAMP 2× Master Mix (New England Biolabs, Ipswich, MA, USA).
2. Colorimetric LAMP uses primers that amplify the target region through elongation via a hairpin structure with stem loops at each end [20, 21]. The primers used are specific for both *C. jejuni* and *C. coli*, amplifying the 16S gene. The complete primer sequences are reported in Table 1.
3. Nuclease-free water.
4. Positive LAMP reaction control: DNA of *C. jejuni* strain (ATCC 33291).

3 Methods

3.1 Sample Preparation

1. Homogenize a total of 25 g of the sample in a stomacher for 1 min with 225 mL of Preston broth (enrichment broth specific for *Campylobacter* spp. detection) (*see* **Note 1**).
2. Incubate the enrichment at 41.5 °C in microaerophilic conditions (with 5% CO_2) for 6 h.

3.2 DNA Extraction

1. After 6 h of enrichment, collect 10 mL of enrichment broth in a 15 mL tube.
2. Centrifuge at 2000 × *g* for 15 min to separate residual matrices.
3. Collect the supernatant and centrifuge it at 10,000 × *g* for 10 min (*see* **Note 2**).
4. Resuspend the pellet in 100 μL of Chelex solution (6%), and transfer it to a 1.5 mL tube (*see* **Notes 3** and **4**).
5. Incubate samples at 56 °C for 20 min and then at 99 °C for 15 min.
6. Centrifuge the tubes at 12,000 × *g* for 5 min.
7. Collect the supernatant, and perform the LAMP reaction (*see* **Note 5**).

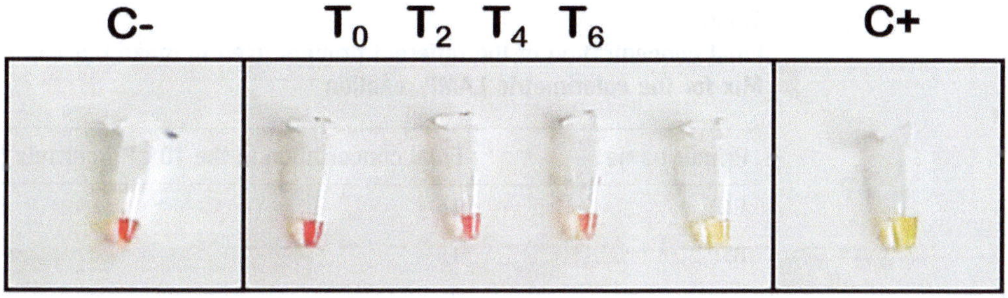

Fig. 1 Colorimetric LAMP result for *Campylobacter* spp. detection; the tube number 1 was positive for *Campylobacter* spp. detection (the color turned from pink to yellow), whereas the tubes from 2 to 6 are negative for *Campylobacter* spp. detection (the color remained pink). (Reproduced from Zendrini et al. [18])

3.3 LAMP Reaction

1. Prepare the LAMP reaction mix using per sample: 12.5 μL of WarmStart Colorimetric LAMP 2× Master Mix (New England Biolabs, Ipswich, MA, USA), 2.5 μL of 10× Primer mix, and 8 μL of nuclease-free water (*see* **Notes 6** and 7).
2. Add 2 μL of the extracted sample to 23 μL of LAMP reaction mix; repeat this point for every sample (*see* **Notes 8**, **9** and **10**).
3. Mix by pipetting and close the tubes.
4. Check that the reaction solution has a bright pink color, required for successful colorimetric LAMP reaction.
5. Put the tubes in the heated plate or water bath for incubation at 60 °C for 45 min (*see* **Note 11**).
6. Remove tubes from the heated plate/water bath and cool on ice to stop the reaction.
7. Examine tubes by the naked eye: positive reactions should have turned yellow, while negative reactions should have remained pink. An example is reported in Fig. 1. Any sample with a color in the spectrum between hot pink and yellow should be considered positive.
8. The result can be photographed at room temperature to document the colorimetric result (*see* **Note 12**).

4 Notes

1. During the sample preparation phase, process always a negative process control consisting only of the Preston broth used for the sample enrichment.
2. At the beginning of the extraction phase, it is possible to store the pellet obtained after the second centrifugation step, at −20 °C until use.

Table 2
Final concentration of the different primers used to make the 10× Primer Mix for the colorimetric LAMP reaction

Primer name	Final concertation in the 10× Primer mix (μM)
FIP	16
BIP	16
F3	2
B3	2
LOOP F	4
LOOP B	4

3. During the extraction phase, process always a negative extraction control (100 μL of Chelex solution 6%).
4. During the extraction phase, process always a positive extraction control (*C. jejuni* dried strain ATCC 33291), following the same protocol as samples.
5. At the end of the extraction phase, it is possible to store the supernatant at −20 °C until use.
6. Before the LAMP reaction phase, prepare the heated plate or water bath in advance to ensure a stable temperature for the whole duration of the LAMP reaction.
7. To prepare LAMP reaction, we suggest to dilute every primer in nuclease-free water to reach the concentration indicated in Table 2, add every primer in equal parts in a tube to prepare the 10× Primer Mix, and then store it at −20 °C until use.
8. In the LAMP reaction phase, the use of 0.1 or 0.2 mL tubes is recommended.
9. During the LAMP reaction phase, process always a negative reaction control (23 μL of LAMP mix with 2 μL of nuclease-free water in place of sample).
10. During the LAMP reaction phase, process always a positive reaction control (23 μL of LAMP mix with 2 μL of DNA of *C. jejuni* strain ATCC 33291).
11. In the LAMP reaction phase, if the color change is not robust or complete (i.e., the color turned from pink to orange), it is recommended to increase the temperature of the heated plate to 65 °C, and leave the tubes for additional 10 min.
12. To document the colorimetric LAMP result, it is possible to keep tubes at room temperature because they will maintain the final color pattern.

Acknowledgments

This research was funded by the Italian Ministry of Health (project ID: RF-2016-02364002).

References

1. Authority EFS and European Centre for Disease Prevention and Control (2021) The European union one health 2019 zoonoses report. EFSA J 19:e06406
2. Facciolà A, Riso R, Avventuroso E, Visalli G, Delia SA, Laganà P (2017) *Campylobacter*: from microbiology to prevention. J Prev Med Hyg 58:E79–E92
3. Man SM (2011) The clinical importance of emerging *Campylobacter* species. Nat Rev Gastroenterol Hepatol 8:669–685
4. Newell DG, Fearnley C (2003) Sources of *Campylobacter* colonization in broiler chickens. Appl Environ Microbiol 69:4343–4351
5. Bryan FL, Doyle MP (1995) Health risks and consequences of *Salmonella* and *Campylobacter jejuni* in raw poultry. J Food Prot 58:326–344
6. Bull SA, Allen VM, Domingue G et al (2006) Sources of *Campylobacter* spp. colonizing housed broiler flocks during rearing. Appl Environ Microbiol 72:645–652
7. Food Standards Agency (n.d.) A microbiological survey of *Campylobacter* contamination in fresh whole UK-produced chilled chickens at retail sale (Y2/3/4) | Food Standards Agency. https://www.food.gov.uk/research/foodborne-disease/a-microbiological-survey-of-campylobacter-contamination-in-fresh-whole-uk-produced-chilled-chickens-at-retail-sale-y234
8. Bodulev OL, Sakharov IY (2020) Isothermal nucleic acid amplification techniques and their use in bioanalysis. Biochem Biokhimiia 85: 147–166
9. Notomi T, Okayama H, Masubuchi H et al (2000) Loop-mediated isothermal amplification of DNA. Nucleic Acids Res 28:E63
10. Yang Q, Domesle KJ, Ge B (2018) Loop-mediated isothermal amplification for *Salmonella* detection in food and feed: current applications and future directions. Foodborne Pathog Dis 15:309–331
11. Maruyama F, Kenzaka T, Yamaguchi N et al (2003) Detection of bacteria carrying the *stx*2 gene by *in situ* Loop-mediated isothermal amplification. Appl Environ Microbiol 69: 5023
12. Kaneko H, Iida T, Aoki K et al (2005) Sensitive and rapid detection of herpes simplex virus and varicella-zoster virus DNA by Loop-mediated isothermal amplification. J Clin Microbiol 43: 3290–3296
13. Enomoto Y, Yoshikawa T, Ihira M et al (2005) Rapid diagnosis of herpes simplex virus infection by a loop-mediated isothermal amplification method. J Clin Microbiol 43:951–955
14. Lau YL, Meganathan P, Sonaimuthu P et al (2010) Specific, sensitive, and rapid diagnosis of active toxoplasmosis by a Loop-mediated isothermal amplification method using blood samples from patients. J Clin Microbiol 48: 3698–3702
15. Liu A, Guan G, Du P et al (2012) Loop-mediated isothermal amplification (LAMP) method based on two species-specific primer sets for the rapid identification of *Chinese Babesia bovis* and *B. bigemina*. Parasitol Int 61:658–663
16. Li Y, Fan P, Zhou S et al (2017) Loop-mediated isothermal amplification (LAMP): a novel rapid detection platform for pathogens. Microb Pathog 107:54–61
17. Quyen TL, Nordentoft S, Vinayaka AC et al (2019) A sensitive, specific and simple Loop-mediated isothermal amplification method for rapid detection of *Campylobacter* spp. in broiler production. Front Microbiol 10:2443
18. Zendrini A, Carta V, Filipello V et al (2021) One-day molecular detection of *Salmonella* and *Campylobacter* in chicken meat: a pilot study. Foods Basel Switz 10
19. EUR-Lex – 32005R2073 – EN – EUR-Lex. Available online: https://eur-lex.europa.eu/legal-content/IT/ALL/?uri=CELEX%3A32005R2073. Accessed on 21 Dec 2023
20. Romero MR, Cook N (2018) A rapid LAMP-based Method for screening poultry samples for *Campylobacter* without enrichment. Front Microbiol 9:409068
21. Romero MR, D'Agostino M, Arias AP et al (2016) An immunomagnetic separation/Loop-mediated isothermal amplification method for rapid direct detection of thermotolerant *Campylobacter* spp. during poultry production. J Appl Microbiol 120:469–477

Chapter 5

Detection of Shiga Toxin-Producing *Escherichia coli* with Loop-Mediated Isothermal Amplification (LAMP) and Bioluminescent Assay in Real Time (BART)

Gabriela Lopez Velasco, Christina Ann Barnes, and Neil Percy

Abstract

Shiga toxin-producing *Escherichia coli* (STEC) are pathogenic microorganisms that can significantly impact public health. Consumption of improperly cooked ground meats, raw milk, and raw vegetables have been associated with outbreaks involving STEC. Methods to evaluate their presence in foods are routinely used in food testing laboratories worldwide to assess their prevalence during the food supply chain and thus prevent the occurrence of consumption of contaminated foods. While applications of traditional microbiological methods and molecular methods involving polymerase chain reaction are available, alternative technologies may become valuable diagnostic tools. Loop-mediated isothermal amplification is a specific and rapid method that enables fast amplification of a target sequence at a constant temperature. In this chapter we have included a revision of the different targets to identify relevant STEC serogroups using LAMP in combination with different reporting technologies to visualize LAMP products. In specific, we describe an optimized LAMP method for identification of STEC in various food matrices targeting Shiga toxin genes utilizing bioluminescence real-time detection to visualize the LAMP products.

Key words STEC, Shiga toxin genes, stx_1, stx_2, Loop-mediated isothermal amplification, Real-time bioluminescence

1 Introduction

1.1 STEC as a Relevant Group of Human Pathogens

Shiga toxin-producing *Escherichia coli* (STEC) are important public health hazards [1]. Just in the USA, over 250,000 illnesses associated to this group of microorganisms were reported [1] and over 8000 confirmed cases were reported in the European Union [2]. STEC can result in an infection of variable severity often characterized with bloody diarrhea and abdominal cramps. Illness may be complicated by hemolytic uremic syndrome (HUS) in kids or thrombotic thrombocytopenic purpura (TTP) in adults.

Alejandro Garrido-Maestu and Alexandre Lamas (eds.), *Foodborne Pathogens: Loop-mediated Isothermal Amplification Protocols*, Methods and Protocols in Food Science, https://doi.org/10.1007/978-1-0716-4526-0_5,

STEC is only one of the four generally recognized groups of *E. coli* causing diarrhea in humans who acquired infections by ingestion of contaminated food or water or other fecal-oral route. The natural reservoir for STEC is ruminant animals, notably cattle in which STEC can occur as normal intestinal flora [3]. STEC is a category of *E. coli* that carries genes that can produce Shiga toxins, *stx1* with four subtypes (a, c, d, and e), and *stx2*, with 15 subtypes (a to o). These can be further grouped by their specific O type. The relevant serotypes may vary geographically, being serotypes O157, O26, O103, O146, O145, O111, and O91 commonly recognized by health and food safety authorities of relevant importance for public health as they are associated with the majority of sever cases in humans [4]. The most common primary sources of STEC outbreaks include raw or undercooked ground meat products, raw milk, and fecal contamination of vegetables [5]. Prevention of infectious diseases requires control measures at all stages in the food chain, which include control of agricultural production and proper separation of animal and field crops, good agricultural practices, and manufacturing and preparation of foods in commercial establishments as well as household kitchens.

Surveillance and routine testing of STEC is performed in food testing and authority surveillance laboratories. For detection of food items, the ISO/TS 13136 [6] and the USDA FSIS MLG5B.05 [7] reference methods include an initial screening for the presence of Shiga toxin genes and/or *eae* genes. The *eae* genes codify the adherence factor, intimin, that enables the adherence of bacteria to gut epithelial cells. Both intimin and Shiga toxin play a critical role in pathogenesis [8]. Commonly, these and other commercial methods validated using reference methods as comparison have relied in the use of polymerase chain reaction (PCR), but as it will be described in this chapter, alternative DNA amplification methodologies that offer alternatives for food testing diagnostics have emerged.

1.2 Loop-Mediated Isothermal Amplification for Detection of Shiga Toxin-Producing E. coli

Loop-mediated isothermal amplification (LAMP) of DNA was described in 2000 as a highly specific, efficient, and rapid DNA amplification method that works under isothermal conditions to detect genetic targets under constant temperature [9]. In addition, because LAMP is capable of recognizing the target to amplify by six distinct sequences, it is expected to amplify the target sequence with high selectivity [9, 10]. Since the publication of LAMP, over 2000 peer review publications have reported their application for the rapid detection of a variety of molecular targets with applications in clinical, animal, and food safety diagnostics [10].

LAMP amplification products can then be visualized by different end point methods such as electrophoresis or immunochromatography or they can be visualized in real time through turbidity, fluorescence, or bioluminescence [11].

For detection of STEC or their specific serotypes, the implementation of LAMP as an amplification method has been reported for their application in food testing (Table 1).

In this chapter the detection of STEC serogroups O26, O45, O103, O111, O121, O145, and O157 using an optimized LAMP assay and a bioluminescent assay in real time is described. The method is based on the amplification of *stx* genes using LAMP and bioluminescence detection. The method relies in the application of Neogen® Molecular Detection System for the rapid and specific detection of Shiga toxin gene (stx_1 and/or stx_2) from Shiga toxin-producing enterohemorrhagic *Escherichia coli* (STEC) in enriched foods [25, 27]. This method was compared for performance to the United States Department of Agriculture Food Safety and Inspection Service Microbiology Laboratory Guidebook, 5C.01 [7] for meat and poultry and to the United States Food and Drug Administration Bacteriological Analytical Manual, Ch. 4A for sprouts, using an unpaired study design [28].

2 Materials and Equipment

- Calibrated micropipettors to handle 20 μL volumes.
- Filtered tips for micropipettors.
- Sterile filtered laboratory blender bags for sample enrichment.
- Incubators capable of maintaining of 41.5 °C.
- Dry bath incubator capable of measuring a temperature of 100 °C.
- Calibrated partial immersion thermometer capable of measuring a temperature of 100 °C.
- 8 × 12 well microtube rack
- Neogen® Molecular Detection System Instrument (including proprietary accessories: lysis rack, capping and decapping tools, speed loading tray, and heat block).
- Neogen® Chill Block (catalogue number MDSCBIN).
- The Neogen® Molecular Detection Software should be installed in a computer for data analysis and LAMP reaction visualization. The software is available at www.neogen.com website and it should be available for download after filling a registration form.

2.1 Reagents

- Buffered Peptone Water (BPW) ISO formulation.
- Neogen Molecular Detection Asay 2—STEC Gene Screen (*stx*) Kit (catalogue number MDA2STX96).
  - Two pouches containing eight tubes of reagent control (RC) lyophilized reagent (*see* **Note 1**).

Table 1
LAMP methods for STEC with potential for application in food diagnostics

Genetic target	Target STEC	Reporting system	Food testing	References
Ecs_2840	*E. coli* O157:H7	Calcein SYBR Green I	Milk samples artificially contaminated with *E. coli* O157:H7	Wang et al. [12]
Shiga-like toxin (*stx2*) gene	*E. coli* O157:H7	CRISPR/Cas12a (fluorescent-labeled ssDNA reporter)	25 g test portions of romaine lettuce concentrated by membrane filtration	So-Young and Se-Wook [13]
Ecs_2840	*E. coli* O157:H7	CRISPR/Cas12a (fluorescent label reporter or visual inspection under UV light)	Milk	Wang et al. [14]
stx₁, *stx₂*, *and eae* genes	STEC serogroup O26, O45, O103, O111, O121, O145, and O157	Real-time turbidity	25 g test portions of ground meat pre-enriched in 225 mL of pre-warmed BPW supplemented with 8 mg/L of vancomycin	Wang et al. [15]
Z3276 gene (encode a putative fimbrial protein -tip adhesin-)	*E. coli* O157:H7	Turbidity by the naked eye (precipitation of magnesium pyrophosphate) and agarose gel electrophoresis	One g test portions of ground beef, homogenized with 9 mL of tryptic soy broth	Ravan et al. [16]
Z3276 gene (encode a putative fimbrial protein -tip adhesin-)	*E. coli* O157:H7	Agarose gel electrophoresis	Fecal material	Yinur et al. [17]
rfb E gene	*E. coli* O157:H7	SYBR Green II and visual detection for turbidity	Not tested in food matrices	Ranjbar et al. [18]
*rfb*E gene	*E. coli* O157:H7	Agarose gel electrophoresis	Raw milk	Deguo Wang and Guicheng Huo [19]
*rfb*E gene and *stx2* gene	*E. coli* O157:H7	Immunochromatography	Milk	Cui et al. [20]
*rfb*E gene	*E. coli* O157:H7	Agarose gel electrophoresis	Not tested in food matrices	Zhao et al. [21]

(continued)

Table 1 (continued)

Genetic target	Target STEC	Reporting system	Food testing	References
O serogroups specific genes - *wzx* (O-unit flippase) or *wzy* (O-antigen polymerase)	STEC serogroup O26, O45, O103, O111, O121, O145, and O157	Real-time turbidity	Ground beef, beef trim, and produce	Wang et al. [22]
*rfb*E gene	*E. coli* O157: H7	Real-time turbidity	Immunomagnetic separation post enrichment of ground beef samples	Qin et al. [23]
O serogroups specific genes - *wzx* (O-unit flippase) or *wzy* (O-antigen polymerase)	STEC serogroup O26, O45, O103, O111, O121, O145, and O157 Improved detection from *Wang et al. (x)*	Real-time turbidity	Not tested in food matrices	Wang [24]
stx_1 and/or stx_2	STEC serogroup O26, O45, O103, O111, O121, O145, and O157	Real-time detection with bioluminescence utilizing photodiode detection(x)	25–375 g test portions of raw beef trim, raw ground pork, raw poultry parts, and sprouts	Rosauer et al. [25]
stx_1 and/or stx_2 and *eae*	STEC serotypes O26, O45, O103, O111, O121, O145, and O157	Real-time detection with bioluminescence utilizing photodiode detection(x)	25 or 375 g test portions of raw beef trim, raw ground pork, raw poultry parts, mechanically separated chicken, spinach, sprouts, and dried cannabis and hemp flowers (10 g)	Rosauer et al. [26]

- Four pouches, each containing three strips of eight orange tubes of STEC Gene Screen (stx) Reagent Tubes containing lyophilized reagents for DNA amplification (*see* **Note 2**).
- One rack holding 12 strips of 8 tubes, with each tube containing 580 μL of a ready-to-use lysis solution (LS) (*see* **Note 3**).

  NOTE: All kit components should be maintained refrigerated (4–8 °C).

3 Method

3.1 Sample Preparation for Enrichment

- Aseptically weigh out the appropriate test portion into a sterile filtered blender bag (*see* **Note 4**).
  - Raw ground beef and trim: 25 g should be suspended in 225 mL of pre-warmed at 41.5 °C BPW ISO.
  - Raw ground beef and trim: 375 g should be suspended in 1125 mL of pre-warmed at 41.5 °C BPW ISO.
  - Raw meat (pork, poultry, lamb, bison): 375 g should be suspended in 1125 mL pre-warmed at 41.5 °C BPW ISO.
  - Leafy produce: 200 g should be suspended in 450 ml of pre-warmed at 41.5 °C BPW ISO.
  - Sprouts: 25 g should be suspended in 225 mL of pre-warmed at 41.5 °C BPW ISO.
- The test portions and corresponding volume of pre-warmed BPW ISO should be homogenized to disperse and break apart clumps. Except for leafy greens and sprouts that should only be agitated to rinse over material but they should not be massaged or homogenized (*see* **Notes 5** and **6**).
- Close the bags.
- Place samples in an incubator at 41.5 °C and incubate the bags for 10–18 h for raw meats and 18–24 h for leafy greens and sprouts.

3.2 Lysis Process

- Place a heating block (provided as an accessory of the Neogen® Molecular Detection System Instrument) into a dry bath incubator.
- Turn on the dry bath incubator and allow the heat block to reach 100 °C ± 1. Verify the heat block temperature with a calibrated partial immersion thermometer (*see* **Note 7**).
- Remove from the refrigerator one lysis solution (LS) tube per sample and two additional LS tubes (one for preparing a negative control and one for preparing a positive control).

- Place the LS tubes in a lysis rack (provided as an accessory of the Neogen® Molecular Detection System Instrument), place the rack in a heat block at 100 °C, and incubate for 30 s to bring the LS tubes at ambient temperature.
- Invert the rack once to wet the wall of the LS tubes with the lysis solution (*see* **Note 8**).
- Open the LS tubes with a capping|decapping lysis tube tool (provided as an accessory of the Neogen® Molecular Detection System Instrument). The caps may be discarded (*see* **Note 9**).
- Remove the enriched sample from the incubator, gently massage the bottom of the enrichment, and collect 20 μL of the enriched sample using a calibrated pipettor and a filtered tip.
- Transfer the 20 μL of the enriched sample to one LS tube by releasing the volume of the side of the tube wall (*see* **Notes 8** and **10**). The LS tubes should not be re-capped.
- Discard the tip and repeat the process for the next sample.
  - For a negative control sample, transfer 20 μL of sterile BPW ISO into an LS tube.
  - For a positive control sample, transfer 20 μL of BPW ISO that has been artificially inoculated with a STEC into a LS tube. It is critical to maintain control of the strains used for positive controls and the manipulation of cultures to prevent cross-contact contamination.
  - Controls may be processed after all test samples have been processed to minimize cross-contact contamination.
- Place the lysis rack containing the LS tubes in a heat block at 100 °C for 15 min ± 1. During this time the LS will change from pink to yellow (*see* **Note 11**).
- Transfer the lysis rack containing the LS tubes into a chill block (provided as an accessory of the Neogen® Molecular Detection System Instrument) at ambient temperature (20–25 °C), and allow to cool down for 5 and up to 10 min. During this time the LS will change from yellow to pink (*see* **Notes 12** and **13**).

3.3 Loop-Mediated Isothermal Amplification

3.3.1 Instrument Preparation

- Turn on the Neogen Molecular Detection Instrument. The instrument will perform a self-check and if all parameters pass, the instrument LED display light will turn orange to warm up and it will turn green once it has reached the proper temperature.
- Launch the Neogen Molecular Detection Software on the computer.
  - Click Set up a new run.
  - Select the number of wells required for test samples and the positive control:

  - Select the icon STX-2.
  - Each individual well requires to be labeled with a sample ID. This can be performed in the Well Detail window using the "Sample" field displayed in the software.
- Select the wells where the reagent control tube and the negative control will be located:
  - Select the icon STX-2 NC to assign the negative control well.
  - Select the icon STX-2 RC to assign the reagent control well.
- Select all the wells and input the lot number of the kit. This can be performed in the Well Detail window using the "Kit Lot Number" field displayed in the software.
- Save the run.
- A green arrow icon (Start button) will display once the run is being saved and the instrument is warmed up at the right temperature.

3.3.2 Preparation of stx Reagent Tubes for DNA Amplification

- Remove from the refrigerator the Neogen Molecular Detection Assay 2—STEC Gene Screen *(stx)* Reagent tube strips needed. Tubes can be cut to desired tube number. A total of one reagent tube is required per each sample and two additional reagent tubes are needed if positive and negative controls are included (*see* **Note 14**).
- Place the Neogen Molecular Detection Assay 2—STEC Gene Screen *(stx)* Reagent tubes in a 8 × 12 microtube rack.
- Remove a Reagent Control tube and place into the microtube rack.
- Use the Capping|Decapping Reagent tube tool to de-cap one Neogen Molecular Detection Assay 2—STEC Gene Screen (*stx*) Reagent Tubes-strip at a time. Discard cap.
- Transfer 20 μL of sample lysate from the upper half of the liquid (avoid precipitate) in the LS tube into corresponding Neogen® Molecular Detection Assay 2—STEC Gene Screen (*stx*) Reagent Tube. Dispense at an angle to avoid disturbing the pellets. Mix by gently pipetting up and down five times (*see* **Note 15**).
- Repeat until individual samples and if prepared, the positive control lysate, have been added to a corresponding Neogen® Molecular Detection Assay 2—STEC Gene Screen (*stx*) Reagent Tube in the strip.
- Re-cap the Reagent Tubes with new caps (provided with the kit).
- To prepare the negative control, transfer 20 μL of the lysed negative control to a Neogen Molecular Detection Assay 2—STEC Gene Screen (*stx*) Reagent Tube and re-cap the tube.

- To prepare the reagent control, transfer 20 μL of the lysed negative control to a Reagent Control tube (single clear tube) and cap the tube.
- Transfer all the tubes in the same order as they were configured in the software run into a Speed loader Tray (provided as an accessory of the Neogen® Molecular Detection System Instrument).
- Click on the Start button in the Neogen Molecular Detection Assay Software and select instrument for use. The selected instrument's lid automatically opens. The instrument should always being opened through the software rather than manually.
- Place the Speed Loader Tray into the Instrument and close the lid to start the assay.

Results are provided within 60 min, although positives may be detected sooner (*see* **Note 16**).

After the assay is complete, remove the Molecular Detection Speed Loader Tray from the Instrument and dispose of the tubes by soaking in a 1–5% (v/v) household bleach in water solution for 1 h and away from the assay preparation area (*see* **Note 17**).

3.3.3 Interpretation of Results

An algorithm in the software will interpret the light output curve resulting from the detection of nucleic acid amplification. Results are analyzed automatically by the software and are color-coded based on the result. A positive or negative result is determined by analysis of several unique curve parameters (Fig. 1). Presumptive positive results are reported in real time, while negative results will be displayed after the run is completed (*see* **Note 18**). A presumptive positive result indicates that the *stx* target has been detected in the samples analyzed and thus the sample is STEC positive (*see* **Note 19**).

4 Notes

1. The reagent control (RC) contains all the lyophilized reagents required for DNA amplification. It also contained an artificial DNA sequence and the required primers to amplify it. The RC is used to verify that the kit components are working, and the kit has been stored properly during shipping and during storage in the laboratory. The reagent control should always result in a positive amplification. If the RC amplification results were negative, this may indicate that the kit components are not working properly. This can be prevented by keeping the kit refrigerated and the pouches properly sealed after they are opened.
2. The STEC Gene Screen (stx) reagent tubes contained all the lyophilized reagents required for amplification of the *stx*1 and/or *stx*2 genes. It is critical to re-close the pouch after opening to prevent exposure of the lyophilized pellet to

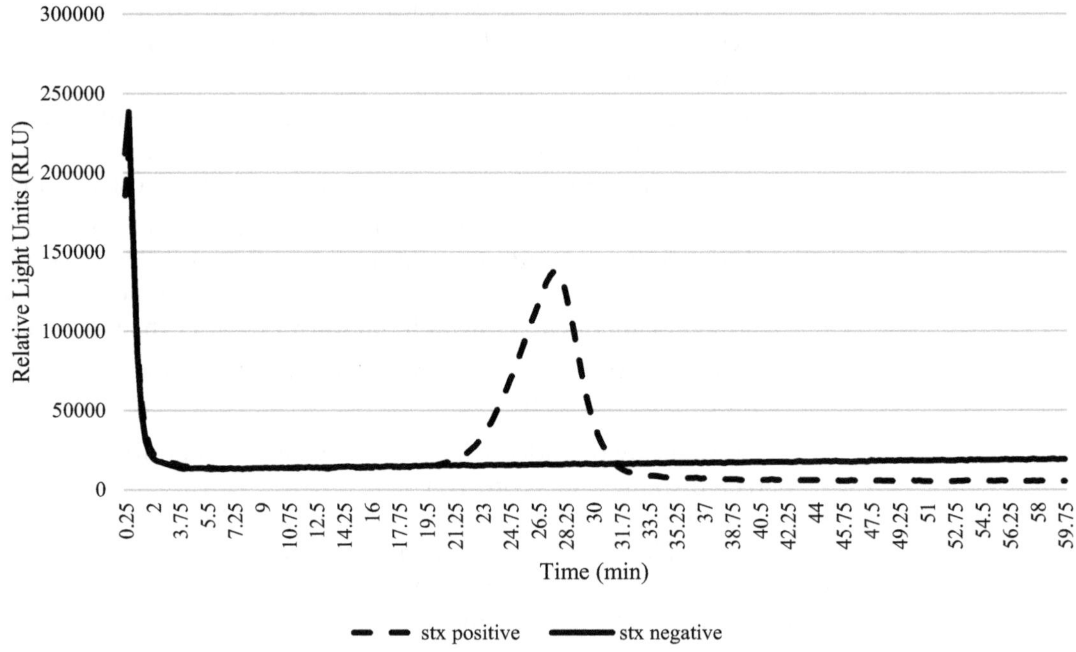

Fig. 1 Raw data obtained from a negative (– Solid line) and a positive (+ dashed line) samples analyzed with LAMP and bioluminescence targeting *stx*. The software displays a red icon with a positive symbol for samples where *stx* was detected and a green icon with a negative symbol for samples where *stx* was not detected. The time and height at which a positive sample sharply peaks is dependent of various factors including the amount of DNA, type of sample etc

humidity. This exposure can cause shrinkage of the lyophilized pellets and can result in inaccurate results. After opening the pouch, the shelf life of the reagent from an opened and resealed pouch is 90 days.

3. The lysis solution (LS) tubes contain a ready-to-use buffer that will enable both the lysis of the microbial cells and the cleaning of the sample. It is important to maintain it refrigerated together with the other kit components.
4. The test portions and corresponding volume of BPW had been optimized for sample enrichment; thus under the conditions described, it is expected that very low levels of STEC grow during incubation at a level where they can be detected downstream. If a different food item or a different sample size is required, it would be important that the laboratory verify performance following appropriate guidelines for method validation and/or method verification.
5. Leafy green vegetables and sprouts should only be agitated to perform an extraction of the microorganisms and ensure the entire leafy material is submerged in the media, but they should not be homogenized in a blender, thoroughly massaged, or macerated as this can interfere with the recovery and growth of STEC.

6. In the case of leafy greens, it is important to ensure that all the sample material is covered by the BPW ISO to expose all the sample surface to the enrichment medium.
7. Verification of the temperature of the heat block with a partial immersion thermometer is essential to ensure optimum lysis conditions; on occasion the display temperature of the dry bath may need to be adjusted to ensure the heat block is at the proper temperature.
8. Inverting the rack containing the LS tubes will wet the wall of the tubes with the LS.
9. The lysis should be done with the LS tubes open; however it is possible to store the lysates for short term in case retesting is necessary, and it is possible to store the caps on a sterile petri dish in case tubes need to be re-capped for storage under refrigeration.
10. The tip should touch the side of the tube wall without immersing it into the LS. Since the tube was inverted and the well of the tube contains LS, the sample will slide down the wall. It is not necessary to mix.
11. Notice that the entire rack should be placed on the heat block to reduce manipulation of tubes. A drop in temperature in the thermometer may be noticed, but this is normal, and temperature should not be adjusted. During this time the sample is lysed. The heat and the contents of the LS will heat and mix the sample.
12. The cooling time is critical to bring the LS to a temperature compatible with the lyophilized reagents used for amplification but also enable the sample to settle; during the cooling time, you will notice that components of the enriched sample and associated debris will deposit on the bottom of the LS tube and the extracted DNA will remain in the upper half of the solution.
13. The DNA will remain in the upper half of the solution but by density will tend to go to the bottom; thus it is critical that the sample is processed for amplification after 5 min of cooling or up to 10 min. If more than 10 min have passed, it is recommended to heat the rack with the LS tubes for 5 min at 100 °C + 1 and cool down for 5 min into the chill block.
14. Inspect that each reagent tube contains a pellet at the bottom; on occasion flicking the strips or individual tubes may be necessary.
15. Most micropipettes contain two stops; for this step, it is important to collect 20 μL of the lysate and deposit them on the top side of the STEC Gene Screen (stx) Reagent Tube and release to the first break. Then, without releasing the break, go to the

bottom of the tube to mix up and down five times. When mixing it is also critical to maintain the micropipette to the first break to prevent foam formation.

16. There are three enzymatic systems within the lyophilized pellets. If the genetic target is present, the DNA polymerase will amplify the region and as a sub-product pyrophosphate is produced. The pyrophosphate is converted by a second enzymatic system to ATP. The ATP is taken by a bioluminescent reporter and generates a light output.
17. To minimize the risk of false positives due to cross-contamination, never open reagent tubes containing amplified DNA. This includes Neogen Molecular Detection Assay 2—STEC Gene Screen (stx) Reagent and Reagent Control Tubes. Always dispose of sealed reagent tubes by soaking in a 1–5% (v/v) household bleach in water solution for 1 h and away from the assay preparation.
18. A negative sample will not give a zero reading as the system and the reagents have a "background" relative light unit (RLU) reading.
19. The definition of STEC may vary geographically, while the presence of *stx* is characteristic of STEC; certain regulatory entities may require to test for the presence of *eae* [7]. A similar approach to the method described in this chapter to determine both genetic targets is also reported by Rosauer et al. [26].

References

1. CDC Center for disease control and prevention. Shiga toxin-producing *E. coli* (STEC) 2018 case definition. Accessed 18 Aug 2024: Shiga Toxin-producing Escherichia coli (STEC) 2018 Case Definition | CDC
2. European Center for Disease Control and Preventiion. Surveillance Report. STEC infection. Annual epidemiological report for 2022. Accessed 18 Aug 24: STEC infection Annual Epidemiological Report 2022 (europa.eu)
3. Hussein HS, Bollinger LM (2005) Prevalence of Shiga toxin-producing *Escherichia coli* in beef cattle. J Food Prot 68(10):2224–2241. https://doi.org/10.1016/j.meatsci.2005.05.012
4. Hunt JM (2010) Shiga toxin-producing *Escherichia coli* (STEC). Clin Lab Med 30(1): 21–45. https://doi.org/10.1016/j.cll.2009.11.001
5. World Health Organization (2024) *E. coli.* http://www.who.int/news-room/fact-sheets. Accessed 29 Aug 2024
6. International Standard Organization. (2012) Microbiology of the food chain—Part 1. Horizontal method for the detection, isolation and characterization of Shiga toxin-producing *Escherichia coli* (STEC). ISO/DIS 13136-1
7. US department of Agriculture (2021). Food Safety and Inspection Service. MLG 5C.01. Detection, isolation and identification of top seven Shiga toxin-producing *Escherichia coli* (STEC) from meat products and carcass and environmental sponges
8. Yang X et al (2020) Genetic diversity of the intimin gene (*eae*) in non-O157 Shiga toxin-producing *Escherichia coli* strains in China. Sci Rep 10:3275. https://doi.org/10.1038/s41598-020-60225-w
9. Notomi T et al (2000) Loop-mediated isothermal amplification of DNA. Nuclei Acids Research 28(12):e63. https://doi.org/10.1093/nar/28.12.e63
10. Soroka M et al (2021) Loop-mediated isothermal amplification (LAMP): the better sibling

of PCR? Cells 10(8):1931. https://doi.org/10.3390/cells10081931

11. Wong YP et al (2018) Loop-mediated isothermal amplification (LAMP): a versatile technique for detection of microorganisms. J Appl Microbiol 124(3):626–643. https://doi.org/10.1111/jam.13647
12. Wang Z et al (2023) Establishment of real-time fluorescence and visual LAMP for rapid detection of *Escherichia coli* O157:H7 and kits construction. Lett Appl Microbiol 76:1–8. https://doi.org/10.1093/lambio/ovad122
13. So-Young L, Se-Wook O (2022) Filtration-based LAMP-CRISPR/Cas12a system for the rapid, sensitivity and visualized detection of *Escherichia coli* O157:H7. Talanta 241: 123186. https://doi.org/10.1016/j.microc.2024.111527
14. Wang Z et al (2024) Establishment of a LAMP-CRISPR/Cas12a for rapid detection of *Escherichia coli* O157:H7 and one-pot detection. Food Microbiol 124:104622. https://doi.org/10.1016/j.fm.2024.104622
15. Wang F et al (2011) Loop-mediated isothermal amplification assays for detecting Shiga toxin-producing *E. coli* in ground beef and human stools. J Clin Microbiol:91–97. https://doi.org/10.1128/JCM.05612-11
16. Ravan H et al (2016) A highly specific and sensitivity loop-mediated isothermal amplification method for the detection of *Escherichia coli* O157:H7. Microb Pathog 91:161–165. https://doi.org/10.1016/j.micpath.2015.12.011
17. Yinur D et al (2023) Loop mediated isothermal amplification as a molecular diagnostic assay: application and evaluation for detection of enterohaemorrhagic *Escherichia coli* (O157: H7). Practical Lab Med 37:e00333. https://doi.org/10.1016/j.plabm.2023.e00333
18. Ranjbar R et al (2016) Visual detection of enterohemorrhagic *Escherichia coli* O157:H7 using Loop-mediated isothermal amplification. Electron Physician 8(6):2576–2585. https://doi.org/10.19082/2576
19. Wang D, Huo G (2012) Rapid detection viable *Escherichia coli* O157:H7 in raw milk using loop-mediated isothermal amplification with aid of ethidium monoazide. Adv Mater Res 343–344:1271–1221. https://doi.org/10.4028/www.scientific.net/AMR.343-344.1217
20. Cui S et al (2024) Visual loop-mediated isothermal amplification (LAMP) assay for rapid on-site detection of *Escherichia coli* O157:H7 in milk products. Food Secur 13:2143. https://doi.org/10.3390/foods13132143
21. Zhao X et al (2013) Rapid detection of viable *Escherichia coli* O157 by coupling propidium monoazide with loop-mediated isothermal amplification. J Microbiol Biotechnol 23(12): 1708–1716. https://doi.org/10.4014/jmb.1306.06003
22. Wang F et al (2012) Rapid and specific detection of *Escherichia coli* serogroups O26, O45, O103, O111, O121, O145, and O157 in ground beef, beef brim, and produce by loop-mediated isothermal amplification. Appl Environ Microbiol 78(8):2727–2736. https://doi.org/10.1128/AEM.07975-11
23. Qin Y et al (2018) Rapid and specific detection of *Escherichia coli* O157:H7 in ground beef using immunomagnetic separation combined with loop-mediated isothermal amplification. Pol J Food Nutr Sci 68(2):115–123. https://doi.org/10.1515/pjfns-2017-0014
24. Wang D (2017) Evaluation and improvement of LAMP assays for detection of *Escherichia coli* serogroups O26, O45, O103, O111, O121, O145, and O157. Afr Health Sci 17(4): 1011–1021
25. Rosauer M et al (2022) Validation of the 3M™ molecular detection assay 2- STEC gene screen (*stx*) for the detection of Shiga toxin gene (*stx1* and/or *stx2*) in additional matrices. AOAC performance tested Method℠ 071903. J AOAC Int 105(4):1126–1135. https://doi.org/10.1093/jaoacint/qsac014
26. Rosauer M et al (2022) Validation of the 3M™ molecular detection assay 2- STEC gene screen (*stx* and *eae*) for the detection of Shiga toxin gene (*stx1* and/or *stx2*) and intimin gene (*eae*): AOAC performance tested Method℠ 071902. J AOAC Int 105(4):1136–1145. https://doi.org/10.1093/jaoacint/qsac015
27. Neogen Corporation (n.d.) Neogen molecular detection assay 2 – STEC Gene Screen (*stx*) instructions for use. Accessed 15 Oct 2024. 7d77464cd4cef307
28. United States Food and Drug Administration (2020). Biological analytical Manual Chapter 4A, Diarrheagenic *Escherichia coli*. Accessed 22 Sep 2024

Chapter 6

LAMP Detection of Pathogenic *Vibrio* spp. in Seafood and Water Samples Using Colori-Fluorometric Dye

Wataru Yamazaki

Abstract

The loop-mediated isothermal amplification (LAMP) assay is a simple, rapid, and highly sensitive test that enables isothermal gene amplification in the range of 60–65 °C, allowing for visual confirmation. The LAMP assay can detect a variety of human, animal, and plant pathogen genes simply by unaided eye observation of color change or fluorescence measurement. It is applicable not only for the identification of isolated bacterial colonies but also for the rapid screening of food, environmental water, and fecal samples.

Key words Colori-fluorometric dye LAMP, Loop-mediated isothermal amplification, *Vibrio cholerae*, *Vibrio parahaemolyticus*

1 Introduction

The loop-mediated isothermal amplification (LAMP) assay is a technique for amplifying and detecting DNA in a short time [6]. While amplification is possible using a total of four primers (two inner primers and two outer primers), the addition of two primers, known as loop primers, increases the number of reaction sites, thereby enhancing amplification speed and detection sensitivity [4]. The LAMP assay exhibits high amplification efficiency and specificity, capable of amplifying target genes 10^9–10^{10}-fold within 15–60 min [3, 10–12]. Initially, the LAMP method relied on turbidity change for detection [3], but in recent years, the use of color dye and fluorescence measurement has gained prominence, enabling visual determination through a clearer color change. Simple fluorescence determination using a portable LED light or real-

The original version of the chapter has been revised. A correction to this chapter can be found at https://doi.org/10.1007/978-1-0716-4526-0_21

Alejandro Garrido-Maestu and Alexandre Lamas (eds.), *Foodborne Pathogens: Loop-mediated Isothermal Amplification Protocols*, Methods and Protocols in Food Science, https://doi.org/10.1007/978-1-0716-4526-0_6,

time fluorescence detection using real-time PCR machines has also become widespread [1, 5].

In the LAMP assay, all processes from amplification to determination occur within a single reaction tube. The absence of a need to open the tube lid reduces the risk of laboratory contamination. Moreover, it serves as a simple, rapid, and inexpensive test method that can be executed using a heat block or water bath in settings lacking specialized measuring equipment [1, 5]. Numerous LAMP primers have been reported for detecting a variety of human, animal, and plant pathogen genes, including *Vibrio* spp. [1–3, 9–12]. LAMP reaction reagents are commercially available and can be homemade since their compositions are publicly available [1–3].

The LAMP assay can be used not only for the identification of isolated bacterial colonies but also for rapid screening detection from various sources such as food, environmental water, and clinical specimens such as feces, swab, and whole blood [1–3, 9–12]. In individuals with gastroenteritis, most *V. cholerae* and *V. parahaemolyticus* isolates from feces exhibit positivity for the cholera toxin gene (*ctx*), thermostable direct hemolysin gene (*tdh*), and TDH-related hemolysin gene (*trh*). Conversely, in seafood and environmental water, most *Vibrio* spp. isolates are negative for these genes. Therefore, employing screening tests proves cost-effective for determining the presence of bacteria carrying these genes in food and environmental water [2, 10, 12]. In case of positive screening test results, the *Vibrio* spp. carrying these genes should be isolated using immunomagnetic bead separation or alternative methods.

In the context of domestic and international regulations regarding the sanitation of fresh seafood, quantitative and qualitative tests for *Vibrio parahaemolyticus* serve as indicators of contamination, prioritizing this approach over the mere presence or absence of pathogenic genes. Given that culture-based tests necessitate several days for result determination, the LAMP assay, with its capability to detect *Vibrio parahaemolyticus* in a species-specific manner, emerges as a valuable alternative [8].

In domestic and international laws and regulations concerning the sanitation of fresh seafood, quantitative and qualitative tests for *Vibrio parahaemolyticus* are used as indicators of contamination, rather than the presence or absence of pathogenic genes. Since these culture-based tests require a number of days to determine the results, the LAMP assay, which detects *Vibrio parahaemolyticus* in a species-specific manner, is also useful [8]. This chapter outlines a rapid screening method for *Vibrio* spp. from seafood and water samples. This chapter presents a rapid screening method for *Vibrio parahaemolyticus* and *Vibrio* spp. virulence factors from seafood and environmental water samples.

2 Materials

2.1 Seafood Samples

1. Fresh seafood samples.
2. Alkaline peptone water (APW) or a culture broth capable of growing *Vibrio* spp.
3. Stomacher (optional).
4. Sterile plastic stomacher bag.
5. Scale.
6. Sterile scissors.
7. Pipette.
8. Filtered tip.
9. 1.5 mL microcentrifuge tube.

2.2 Environmental Water Samples

1. Environmental water samples.
2. 2× (double concentration) alkaline peptone water (APW) or 2× enrichment broth capable of culturing *Vibrio* spp. (*see* **Note 1**)
3. A flask of sufficient volume to hold a mixture of environmental water and 2x enrichment broth.
4. Graduated cylinder.
5. Pipette.
6. Filtered tip.
7. 1.5 mL microcentrifuge tube.

2.3 DNA Template

1. Heat block (for 95–100 °C).
2. Vortex mixer.
3. High-speed centrifuge for 1.5 mL microcentrifuge tubes (900–20,000 × *g*).
4. 1.5 mL microcentrifuge tubes.
5. NaOH (sodium hydroxide; 1 mol L^{-1}), stored at room temperature. Diluted 40-fold with sterile distilled water and adjusted to 25 mmol L^{-1}, and store at −20 °C or room temperature until required.
6. Tris-HCl buffer (hydrochloric acid; pH 7.5, 1 mol L^{-1}), stored at room temperature.

2.4 LAMP Primer Sets

The sequences of each primer set that specifically detects the *ctx*, *tdh*, and thermolabile hemolysin gene (*tlh*) are shown in Tables 1, 2, and 3. The sequences of the primer sets for detecting the *trh*1 and *trh*2 at once are shown in Table 4. The LAMP primer mixtures required for each reaction are shown in Table 5 (*see* **Note 2**). For simultaneous screening of two genes of particular clinical importance, the *ctx* and *tdh*, the two LAMP primer sets can be mixed in

Table 1
LAMP primers for the *ctx* detection

Primer	Sequence (5′ to 3′)
CtxA-FIP	TCT GTC CTC TTG GCA TAA GAC GCA GAT TCT AGA CCT CCT G
CtxA-BIP	TCA ACC TTT ATG ATC ATG CAA GAG GCT CAA ACT AAT TGA GGT GGA A
CtxA-F3	GCA AAT GAT GAT AAG TTA TAT CGG
CtxA-B3	GMC CAG ACA ATA TAG TTT GAC C
CtxA-LF	CAC CTG ACT GCT TTA TTT CA
CtxA-LB	AAC TCA GAC GGG ATT TGT TAG G

Amplification is performed at 65 °C for 30 min [9]

Table 2
LAMP primers for the *tdh* detection

Primer	Sequence (5′ to 3′)
Tdh-FIP	GTA CCT GAC GTT GTG AAT ACT GAT TGT CTC TGA CTT TTG GAC AAA C
Tdh -BIP	TGA CAT CCT ACA TGA CTG TGA ACA CTT ATA GCC AGA CAC CGC
Tdh -F3	AGA TAT TGT TTG TTG TTC GAG AT
Tdh -B3	AAC ACA GCA GAA TGA CCG
Tdh -LF	GTA CGG TTT TCT TTT TAC ATT ACG
Tdh -LB	AAG ACT ATA CAA TGG CAG CG

Amplification is performed at 65 °C for 30 min [11]

Table 3
LAMP primers for the *trh* detection

Primer	Sequence (5′ to 3′)
Trh1-FIP Trh2-FIP	TAT AGG CGC TTA ACC ATT TTG ATA AAA GCG TTC ACG GTC AAT TAT AGG CGC TTA ACC ACT TTG AAC AAA GAT GTA TAC GGT CAA T
Trh1 -BIP Trh2 -BIP	TCT GGC TAT AAA GAT GGC CTT TCA AGA ATA GTT CTG ATT TAG GCT TG CTT TCT GGT TAT AAA CAT GGT ACT TCT ATA AAA GTC TTG GTT TAG GCT TG
Trh -F3	GGA GGA CTA TTG GAC AAA C
Trh -B3	TGC CGC TCT CAT ATG C
Trh1 -LF Trh2 -LF	GCC TGA AGT CGT GAA AAT AG ACC CGC AGT TGT GAA AAC CG
Trh -LB	CGG TCT TCA CAA AAT CAG AG

Amplification is performed at 63 °C for 30 min [2]

Table 4
LAMP primers for the *tlh* detection

Primer	Sequence (5′ to 3′)
Tlh-FIP	ATG TTT TTA AAT GAA ACG GAG CTC CGG CAA AAA ACG AAG ATG GT
Tlh -BIP	ACG TCG CAA AAC GTT ATC CGG CGA AGA ACG TAA TGT CTG
Tlh -F3	AGC TAC TCG AAA GAT GAT CC
Tlh -B3	GGT TGT ATG AGA AGC GAT TG
Tlh -LF	ACC AGT AGC CGT CAA TG
Tlh -LB	TTA GAT TTG GCG AAC GAG A

Amplification is performed at 65 °C for 30 min [8]

Table 5
Composition of LAMP reaction reagent

Primer mix (for the *ctx, tdh*, or *tlh* genes)		
Component	**Amount**	**(μL)**
100 μmol L^{-1} FIP	40	μL
100 μmol L^{-1} BIP	40	μL
100 μmol L^{-1} LF	20	μL
100 μmol L^{-1} LB	20	μL
100 μmol L^{-1} F3	5	μL
100 μmol L^{-1} B3	5	μL
Total (for 100 runs)	130	μL

Primer mix (for *trh* gene)		
Component	**Amount**	**(μL)**
100 μmol L^{-1} Trh1-FIP	40	μL
100 μmol L^{-1} Trh2-FIP	40	μL
100 μmol L^{-1} Trh1-BIP	40	μL
100 μmol L^{-1} Trh2-BIP	40	μL
100 μmol L^{-1} Trh1-LF	20	μL
100 μmol L^{-1} Trh2-LF	20	μL
100 μmol L^{-1} Trh-LB	20	μL
100 μmol L^{-1} Trh-F3	5	μL
100 μmol L^{-1} Trh-B3	5	μL
Total (for 100 runs)	230	μL

(continued)

Table 5
(continued)

CFI (Colori-fluorometric indicator)		
Component	**Amount**	**(μL)**
Distilled water	696.5	μL
10 mmol L^{-1} HNB (hydroxynaphthol blue)[a]	300	μL
Gel green (×10,000)	3.5	μL
Total (for 1000 runs)	1000	μL
[a]Distilled water: HNB powder = 1 mL: 6 mg		

2× LAMP reaction mix (Homemade)		
Component	**Amount**	**(mL)**
Distilled water	5.85	mL
5 mol L^{-1} betaine	4	mL
1 mol L^{-1} Tris-HCl (pH 8.8)	0.5	mL
1 mol L^{-1} potassium chloride (KCl)	0.25	mL
1 mol L^{-1} $(NH_4)_2SO_4$	0.25	mL
1 mol L^{-1} $MgSO_4$	0.2	mL
100 mmol L^{-1} dATP	0.35	mL
100 mmol L^{-1} dCTP	0.35	mL
100 mmol L^{-1} dGTP	0.35	mL
100 mmol L^{-1} dTTP	0.35	mL
Tween 20	0.5	mL
Total (for 1000 runs)	12.5	mL

LAMP reaction mixture for individual detection of *ctx*, *tdh*, or *tlh* genes		
Component	**Amount**	**(μL)**
2× LAMP master mix	12.5	μL
Distilled water	8.2	μL
Primer mix (for *ctx*, *tdh*, or *tlh* gene)	1.3	μL
CFI	1	μL
Bst 2.0 DNA polymerase (8 units/μL)	1	μL
Total	24	μL

(continued)

Table 5
(continued)

LAMP reaction mixture for simultaneous detection of *ctx* and *tdh* genes		
Component	**Amount**	**(μL)**
2× LAMP master mix	12.5	μL
Distilled water	6.9	μL
Primer mix for *ctx* gene	1.3	μL
Primer mix for *tdh* gene	1.3	μL
CFI	1	μL
Bst 2.0 DNA polymerase (8 units/μL)	1	μL
Total	24	μL

LAMP reaction mixture for the *trh* gene detection		
Component	**Amount**	**(μL)**
2× LAMP master mix	12.5	μL
Distilled water	7.2	μL
Primer mix for *trh* gene	2.3	μL
CFI	1	μL
Bst 2.0 DNA polymerase (8 units/μL)	1	μL
Total	24	μL

one LAMP reaction tube. In that case, reduce the sterile distilled water by the amount of primer mix (1.3 μL) to be added and adjust the total master mix volume to 24 μL (Table 5). Primer mixtures are prepared in 1.5 mL screw-cap tubes and stored at −20 °C until use. If not to be used for an extended period, store at −80 °C.

2.5 LAMP Reaction

1. Commercially available LAMP reagent kits (e.g., Loopamp DNA amplification kit) (Eiken Chemical, Tokyo, Japan) or homemade LAMP reagents as shown in Table 5.
2. Colori-fluorometric indicator (CFI; see Ref. [1]) Composition as shown in Table 5
3. Heat block or water bath for color change confirmation (used at isothermal temperatures of 63 °C or 65 °C).
4. (Optional.) Fluorescence measurement device such as a portable LED light emitting 470 nm or a real-time PCR machine for real-time fluorescence monitoring (used at isothermal temperatures of 63 °C or 65 °C)

5. Vortex mixer.
6. Tabletop centrifuge.
7. Pipette.
8. Filtered tip.
9. 1.5 mL microcentrifuge tube
10. 0.2 mL clear plastic tubes for color change observation by unaided eye.

3 Methods

Debris derived from seafood or environmental water may inhibit the LAMP reaction and should be removed. The use of commercial DNA extraction kits is reliable, but expensive. In this chapter, a simple and inexpensive method of LAMP amplification of crude DNA extracted only by centrifugation three times and lysis of the bacteria with NaOH solution, and neutralization with Tris-HCl buffer, without using a commercial DNA extraction kit, will be presented.

3.1 DNA Extraction from Seafood Sample

1. Prepare a seafood sample homogenate adjusted to a 10% concentration by placing APW or other enrichment broth in a plastic stomacher bag (e.g., add 225 mL of APW for 25 g of seafood).
2. Homogenize the stomacher bag containing the mixture of seafood and enrichment broth by soft hand massage or homogenize in a stomacher for 15–30 s (*see* **Note 3**).
3. Incubate the stomacher bag containing the mixture of seafood and enrichment broth at 35–37 °C for 6–24 h.
4. Transfer 1 mL of the enrichment broth cultures to a 1.5 mL microcentrifuge tube and centrifuge at 900–1,000 *g* for 1 min (*see* **Note 4**).
5. Transfer 500 μL of the supernatant to a new 1.5 mL microcentrifuge tube.
6. Centrifuge at 10,000 × *g* for 5 min and remove the supernatant (*see* **Note 5**).
7. Suspend the pellet in 100 μL of NaOH (25 mmol L^{-1}) (*see* **Note 6**).
8. Vortex to dissolve the pellet and heat at 95–100 °C for 5–10 min.
9. After flush, neutralize the solution by adding 8 μL of Tris-HCl buffer (1 mol L^{-1}, pH 7.5).
10. Centrifuge at 20,000 *g* for 5 min.
11. Use 1 μL of the supernatant as template DNA for the LAMP reaction.

3.2 DNA Extraction from Water Sample

1. Place APW or other enrichment broth adjusted to 2× concentration in a glass flask and add an equal volume of environmental water sample (e.g., add 500 mL of 2× APW to 500 mL of environmental water) (*see* **Note 1**).
2. Agitate the glass flask containing the mixture of environmental water and concentrated broth.
3. Incubate the glass flask containing the mixture of environmental water and enrichment broth at 35–37 °C for 6–24 h.
4. Transfer 1 mL of the enrichment broth cultures to a 1.5 mL microcentrifuge tube and centrifuge at 900–1,000 *g* for 1 min (*see* **Note 4**).
5. Transfer 500 μL of the supernatant to a new 1.5 mL microcentrifuge tube.
6. Centrifuge at 10,000 × *g* for 5 min and remove the supernatant (*see* **Note 5**).
7. Suspend the pellet in 100 μL of NaOH (25 mmol L^{-1}) (*see* **Note 6**).
8. Vortex to dissolve the pellet and heat at 95–100 °C for 5–10 min.
9. After flush, neutralize the solution by adding 8 μL of Tris-HCl buffer (1 mol L^{-1}, pH 7.5).
10. Centrifuge at 20,000 × *g* for 5 min.
11. Use 1 μL of the supernatant as template DNA for the LAMP reaction.

Another method is to trap bacteria including *Vibrio* spp. in environmental water on a 0.45 μm diameter membrane filter and incubate the filter in a 1× enrichment broth for *Vibrio* spp. detection.

3.3 LAMP Reaction and Confirmation (See Note 7)

1. Thaw the LAMP reagent at room temperature and store it on ice.
2. While operating on ice, prepare a master mix in a 1.5 mL or 0.5 mL microcentrifuge tube.
3. Mix the reagents of the composition shown in Table 5.
4. Dispense 24 μL of the master mix into a LAMP reaction tube.
5. Add 1 μL of DNA template to 24 μL of master mix. To prevent contamination caused by the DNA template being sucked into the pipette, pipette only a minimum number of times (1–2 times) and with minimal force.
6. Close the cap of the tube, mix by vortexing, and centrifuge for 2–3 s (*see* **Note 8**).
7. Incubate the mixture at 63 °C or 65 °C for 30 min using a heat block, water bath, or real-time measurement device.

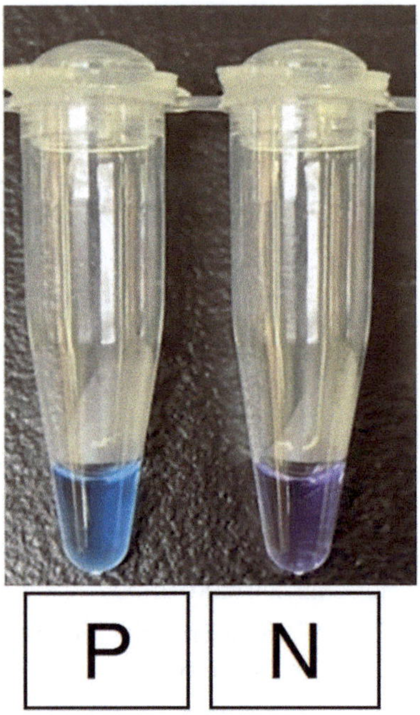

Fig. 1 Visual detection of CT-producing *V. cholerae* by color change; tube P, positive sample; tube N, negative sample

8. To terminate the reaction, inactivate the polymerase at 80 °C for 5 min or 95 °C for 2 min (*see* **Note 9**).
9. The reaction is considered positive when a color change form violet to sky blue is visible to the unaided eye in the LAMP reaction tube or when the fluorescence is automatically calculated as positive in a real-time measurement device or from the endpoint observation by using an LED light emitting 470 nm (*see* **Note 10** and Fig. 1).

4 Notes

1. A 2× (double concentration) enrichment broth is prepared by reducing the amount of distilled water by one-half when the enrichment broth powder is dissolved in distilled water.
2. Use sufficiently purified LAMP primers for rapid and stable gene amplification. All LAMP primers used should be ordered with at least sequence grade and should be manufactured by a reliable supplier with high quality.
3. Food components such as organic and phenolic compounds, glycogen, fats, and calcium ions have been previously reported to inhibit DNA polymerase activity [7, 8]. Prolonged stomaching procedures appear to exacerbate the release of inhibitory

factors from the seafood sample; therefore, stomaching procedure should be no longer than 30 s. Light hand massaging is preferable due to the low release of inhibitory factors. *Vibrio* spp. contamination of seafood is surface contamination, not internal to the edible portion. Therefore, gentle conditions of stomaching or hand massaging are sufficient to release *Vibrio* spp. adhering to the surface of the seafood into the enrichment broth.

4. The first centrifugation at 900–1000 *g* for 1 min is performed to remove large debris in the sample, which reduces the influence of inhibitors derived from seafood and environmental water. Because the centrifugation conditions are so mild, many bacteria remain suspended in the enrichment culture broth without precipitation.
5. The second centrifugation at 10,000 *g* for 5 min is useful for concentrating bacterial cells in the enrichment broth cultures. If seafood-derived components are visually observed adhering to the wall of the microcentrifuge tube, they should be removed using a sterile cotton swab. This is because if not removed, they could eventually remain in the DNA template and inhibit the LAMP reaction.
6. This can be avoided by NaOH heat treatment, which may lyse and inactivate one or more unidentified inhibitory factors in the bacterial cells during the LAMP reaction.
7. Given the high sensitivity of LAMP assays that synthesize large amounts of DNA, amplification of even a small amount of contaminating product in the reaction tube can lead to false-positive results. This type of contamination can be avoided by thorough zoning, i.e., sample and reagent preparations on different clean benches with different pipets and filtered tips.
8. Avoid large bubbles when mixing the master mix and DNA template solution, as large bubbles in the solution can interfere with real-time measurements and cause false results. If large bubbles are present, tap the tube approximately 2–3 times and spin down the solution to remove the bubbles.
9. If the inactivation process at 80 °C for 5 min or at 95 °C for 2 min is omitted, a visual assessment with the unaided eye should be performed quickly after the 30 min amplification. Failure to make a quick assessment may result in the observation of nonspecific positive reactions in negative samples due to residual heat in the LAMP reaction tube.
10. When using a real-time PCR machine for LAMP measurement, choose 6′-FAM for detection, and no quencher setting is required. Thirty cycles of 65 °C (or 63 °C) for 1 s denaturation followed by 65 °C (or 63 °C) for 59 s annealing and extension are set, which allows for isothermal amplification at 65 °C

(or 63 °C) for 30 min. After this amplification, the LAMP reaction is terminated by enzyme inactivation treatment at 80 °C for 5 min or 95 °C for 2 min to prevent nonspecific amplification.

11. Do not open the cap of a used LAMP reaction tube. Contamination of amplified LAMP products from other samples may lead to contamination of the testing area. Keep used reaction tubes with the caps completely closed and dispose of the reaction tubes by incineration or by double packaging in a sealable plastic bag. To prevent the spread of amplified LAMP product, do not autoclave upon disposal.

References

1. Hayashida K, Kajino K, Hachaambwa L, Namangala B, Sugimoto C (2015) Direct blood dry LAMP: a rapid, stable, and easy diagnostic tool for human African Trypanosomiasis. PLoS Negl Trop Dis 9(3):e0003578. https://doi.org/10.1371/journal.pntd.0003578
2. Malcolm TTH, Cheah YK, Radzi CWJWM, Kasim FA, Kantilal HK, John TYH, Jaime Martinez-Urtaza J, Nakaguchi Y, Nishibuchi M, Son R (2015) Detection and quantification of pathogenic *Vibrio parahaemolyticus* in shellfish by using multiplex PCR and Loop-mediated isothermal amplification assay. Food Control 47:664–671. https://doi.org/10.1016/j.foodcont.2014.08.010
3. Mori Y, Nagamine K, Tomita N, Notomi T (2001) Detection of Loop-mediated isothermal amplification reaction by turbidity derived from magnesium pyrophosphate formation. Biochem Biophys Res Commun 289(1):150–154. https://doi.org/10.1006/bbrc.2001.5921
4. Nagamine K, Hase T, Notomi T (2002) Accelerated reaction by Loop-mediated isothermal amplification using loop primers. Mol Cell Probes 16(3):223–229. https://doi.org/10.1006/mcpr.2002.0415
5. Ngan MT et al (2023) Development of a highly sensitive point-of-care test for African swine fever that combines EZ-fast DNA extraction with LAMP detection: evaluation using naturally infected swine whole blood samples from Vietnam. Vet Med Sci 9(3):1226–1233. https://doi.org/10.1002/vms3.1124
6. Notomi T, Okayama H, Masubuchi H, Yonekawa T, Watanabe K, Amino N, Hase T (2000) Loop-mediated isothermal amplification of DNA. Nucleic Acids Res 28(12):E63. https://doi.org/10.1093/nar/28.12.e63
7. Schrader C, Schielke A, Ellerbroek L, Johne R (2012) PCR inhibitors – occurrence, properties and removal. J Appl Microbiol 113(5):1014–1026. https://doi.org/10.1111/j.1365-2672.2012.05384.x
8. Wilson IG (1997) Inhibition and facilitation of nucleic acid amplification. Appl Environ Microbiol 63(10):3741–3751. https://doi.org/10.1128/aem.63.10.3741-3751.1997
9. Yamazaki W, Ishibashi M, Kawahara R, Inoue K (2008) Development of a Loop-mediated isothermal amplification assay for sensitive and rapid detection of *Vibrio parahaemolyticus*. BMC Microbiol 8:163. https://doi.org/10.1186/1471-2180-8-163
10. Yamazaki W, Seto K, Taguchi M, Ishibashi M, Inoue K (2008) Sensitive and rapid detection of cholera toxin-producing *Vibrio cholerae* using a Loop-mediated isothermal amplification. BMC Microbiol 8:94. https://doi.org/10.1186/1471-2180-8-94
11. Yamazaki W, Taguchi M, Kawai T, Kawatsu K, Sakata J, Inoue K, Misawa N (2009) Comparison of Loop-mediated isothermal amplification assay and conventional culture methods for detection of *Campylobacter jejuni* and *Campylobacter coli* in naturally contaminated chicken meat samples. Appl Environ Microbiol 75(6):1597–1603. https://doi.org/10.1128/AEM.02004-08
12. Yamazaki W, Kumeda Y, Misawa N, Nakaguchi Y, Nishibuchi M (2010) Development of a Loop-mediated isothermal amplification assay for sensitive and rapid detection of the *tdh* and *trh* genes of *Vibrio parahaemolyticus* and related *vibrio* species. Appl Environ Microbiol 76(3):820–828. https://doi.org/10.1128/AEM.02284-09

Chapter 7

Loop-Mediated Isothermal Amplification Assay (LAMP) for Detection of *Staphylococcus aureus* in Meat

Gunimala Chakraborty, Juliet Mohanraj, Masmarika Mohan, and Rajeshwari Vittal

Abstract

Loop-Mediated Isothermal Amplification (LAMP)-based diagnostic test can identify *Staphylococcus aureus* (*S. aureus*) utilizing two target genes, *femA* and *arcC*. The optimal reaction temperature for the *femA* and *arcC* genes was determined to be 65 °C and 60 °C, respectively. The technique specifically amplifies DNA from *S. aureus* only, without cross-reactivity with other similar bacterial species, and demonstrated a 100-fold better sensitivity than PCR. Additionally, when tested in both intact and enriched conditions, the LAMP assay detected the pathogen in food samples, particularly meat and dairy products. Thirteen samples were found to be positive for *S. aureus*, with LAMP revealing a higher number of positive samples than PCR.

Key words LAMP, *Staphylococcus aureus*, *arcC*, *femA*, Genes

1 Introduction

Food poisoning caused by *Staphylococcus aureus* is one of the most commercially significant foodborne infections and a serious concern for public health programs across the world [1, 2]. *S. aureus* is estimated to be present in 15.7% of packaged meat and meat products [3]. The pathogen can be identified using various methods, including microbiological, biochemical, molecular, and immunological testing.

For *S. aureus* identification, polymerase chain reaction (PCR)-based approaches targeting the enterotoxins genes (sea, seg, seh, sei) and 23S rDNA have been developed [3, 4, 9]. Similarly, enzyme-linked immunosorbent test (ELISA) and enzyme-linked fluorescence assay (ELFA) have been utilized, but their sensitivity and specificity are lacking [5]. These methods are expensive and require expert workers. As a result, cost-effective, specific, and sensitive diagnostic methods are necessary for the quick

Alejandro Garrido-Maestu and Alexandre Lamas (eds.), *Foodborne Pathogens: Loop-mediated Isothermal Amplification Protocols*, Methods and Protocols in Food Science, https://doi.org/10.1007/978-1-0716-4526-0_7,

identification of *S. aureus*. Since the year 2000, loop-mediated isothermal amplification (LAMP), a nucleic acid amplification technology, has found extensive application in detecting infectious illnesses, genetic disorders, and genetic traits. This approach has eliminated many of the drawbacks of previous molecular technologies. The LAMP approach amplifies nucleic acids in isothermal conditions without the need for thermal cyclers, making it suitable for resource-constrained laboratories in poor countries [6, 10–12].

In this study, we describe a LAMP assay that targets the genes *femA* and *arcC* genes in pathogenic *S. aureus*. We developed a LAMP assay that targets *femA* and *acrC*, which could potentially be utilized to quickly detect and screen *Staphylococcus aureus* in food samples. The assay's sensitivity is 100 times greater than that of conventional PCR (Fig. 1). The assay's specificity is accurate for

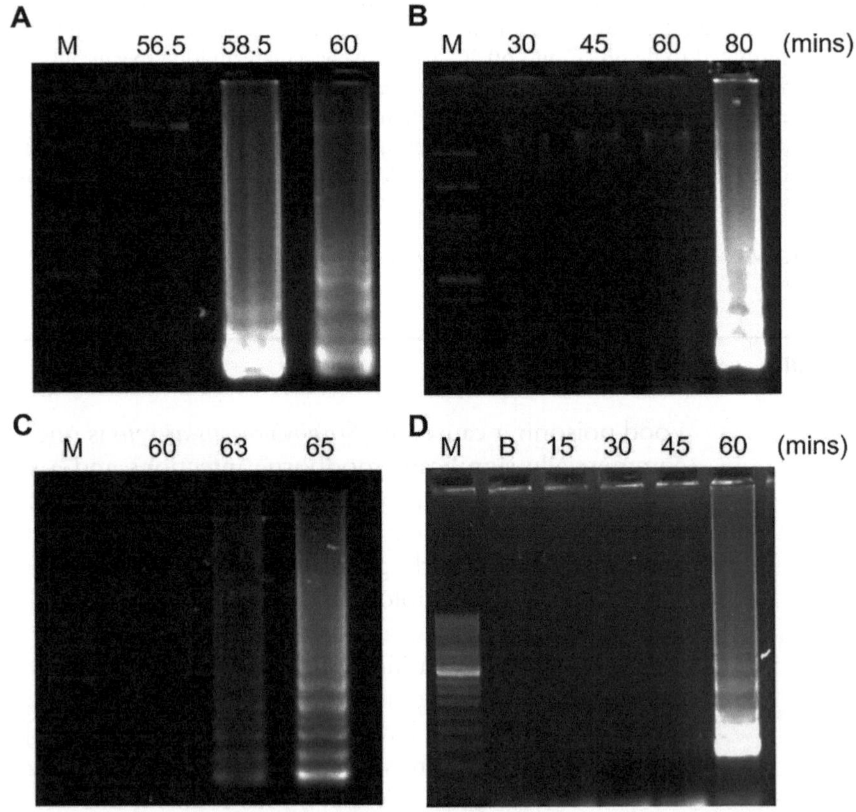

Fig. 1 Optimization of temperature (°C) and time for LAMP reaction for *arcC* and *femA* genes. (**a**) Temperature for *femA*, lane M, molecular marker 100 bp; lane 1, 56.5 min; lane 2, 58.5 min; and lane 3, 60 min. (**b**) time for *femA*, lane M, molecular marker 100 bp; lane 1, 30 min; lane 2, 45 min; lane 3, 60 min, and lane 4, 60 min. (**c**) Temperature for *arcC*, lane M, molecular marker 100 bp; lane 1, 60 min; lane 2, 63 min; and lane 3, 65 min. (**d**) Time for *arcC*, lane M, molecular marker 100 bp; lane 1, 15 min; lane 2, 30 min; lane 3, 45 min; and lane 4, 60 min. All the products were run on 2.0% agarose gel and stained with ethidium bromide

Table 1
Specificity of LAMP reaction for *staphylococcus aureus*

Target organisms	*arc C*	*femA*
Staphylococcus aureus	+ve	+ve
Staphylococcus epidermidis	−ve	−ve
Micrococcus luteus	−ve	−ve
Listeria monocytogenes	−ve	−ve
Bacillus cereus	−ve	−ve
Salmonella Typhimurium	−ve	−ve
Escherichia coli	−ve	−ve

pathogen detection (Table 1). The developed assay was shown to be extremely specific for *Staphylococcus aureus.*

2 Materials

Prepare the reaction mixture using molecular grade water and perform the LAMP assay in laminar air flow to avoid contamination and false positive.

2.1 Bacterial DNA Extraction

1. *Staphylococcus aureus* strain (ATCC 29213) (*see* **Note 1**).
2. Luria-Bertani broth (LB) (HiMedia Laboratories Pvt. Ltd., India) (*see* **Note 2**).

2.2 Primer Construction and Optimization of LAMP

1. Primer expolorerV4 (http://primerexplorer.jp/elamp4.0.0/index.html) software.
2. Two inner primers (FIP and BIP).
3. Two outer primers (F3 and B3).
4. Target genes *femA* (NC_007795.1) and *arcC* (BX571856.1).
5. Loopamp, DNA Amplification kit, (Eiken Chemical Co., LTD., Japan).
6. 2× reaction mixture (40 mM TrisHCl, 20 mM KCl, 16 mM $MgSO_4$, 20 mM $(NH_4)_2SO_4$, 0.2% tween 20, 1.6 M betaine. 2.8 mM dNTPs each)
7. 0.5 μL of *Bst* DNA polymerase
8. 1 μL of target DNA
9. Distilled water.
10. 2% agarose gel and Gel Doc system (Bio-Rad, USA).

2.3 Specificity of LAMP Assay

1. *Staphylococcus aureus.*
2. *Epidermidis* (ATCC 12228).
3. *Micrococcus luteus* (ATCC 9341).
4. *Bacillus cereus* (ATCC 10876).
5. Salmonella Typhimurium (ATCC 14028).
6. *Escherichia coli* (ATCC 8739).
7. *Listeria monocytogenes* (ATCC 35152).

2.4 Sensitivity of LAMP Assay

1. 5 μL of dNTP mixture (2.5 mM each of dATP, dCTP, dGTP, and dTTP)
2. 10× Gene *Taq* universal buffer
3. 0.5 μL of *Taq* polymerase (5 U μL^{-1})
4. 5.0 μL (10 μM) of each primer (forward primer and reverse primer)
5. 5.0 μL of distilled water
6. 1.0 μL of DNA
7. 2% Agarose gel and a Gel Doc system.

2.5 Screening of Staphylococcus aureus from Food Samples

1. The food samples (milk and dairy products, meat samples).
2. Sterile 2.0 mL microcentrifuge tubes (Tarsons Pvt. Ltd., India).
3. 1 mL Butterfield's phosphate-buffered dilution water
4. Baird-Parker agar (BPA) plates and Luria-Bertani broth (LB) plates.
5. Protein A and bound coagulase detection kit (HiMedia, India).
6. 9 mL tryptic soya broth and 1% triton-100.

3 Methods

3.1 Bacterial DNA Extraction

1. Use the *Staphylococcus aureus* strain for assay optimization.
2. Inoculate *Staphylococcus aureus* in Luria-Bertani broth (LB) and incubate at 37 °C overnight with continuous shaking.
3. Extract the genomic DNA using the cetyltrimethylammonium bromide (CTAB) technique [7].

3.2 Optimization of LAMP Conditions

1. Design the LAMP primers based on the sequence information of the target genes *femA* (NC_007795.1) and *arcC* (BX571856.1) using the primer expolorerV4 software (*see* **Note 3**).
2. Perform the LAMP assay using the Loop-amp, DNA Amplification kit (*see* **Note 4**).

3. Add 1 μL each of inner primers (40 pmol each), 0.5 μL of F3 and B3, 6.25 μL of 2× reaction mixture (40 mM TrisHCl, 20 mM KCl, 16 mM $MgSO_4$, 20 mM $(NH_4)_2SO_4$, 0.2% tween 20, 1.6 M betaine. 2.8 mM dNTPs each), 0.5 μL of *Bst* DNA polymerase, and 1 μL of target DNA, and make up the volume to 12.5 μL with distilled water.
4. Optimize the reaction temperature at isothermal conditions ranging from 58 °C to 65 °C (Fig. 1).
5. Optimize the reaction time of the LAMP assay between 15 and 80 min at the optimal temperature, with a 15-mi interval between each time point (Fig. 1).
6. Perform the reaction at isothermal conditions in a dry bath and terminate the reaction by inactivating the enzyme activity at 80 °C for 10 min.
7. For visualization, run gel electrophoresis using a 2% agarose gel and visualize the results using a Gel Doc system.

3.3 Specificity and Sensitivity

1. To determine specificity, perform a LAMP assay using isolated DNA from *S. aureus* and other closely similar gram-positive organisms such as *S. epidermidis*, *Micrococcus luteus*, *Bacillus cereus*, and other food pathogens like Salmonella Typhimurium, *Escherichia coli, and Listeria monocytogenes* (Table 1).
2. To determine the minimum concentration of DNA, start with an initial DNA concentration of 150 ng/μL and conduct the LAMP assay with a tenfold serial dilution of the DNA samples (Fig. 2).
3. To carry out PCR reaction, add 5 μL of dNTP mixture (2.5 mM each of dATP, dCTP, dGTP, and dTTP), 10× Gene *Taq* universal buffer, 0.5 μL of *Taq* polymerase (5 U μL^{-1}), 5.0 μL (10 μM) of each primer (target genes: *arcC* and *femA)*:

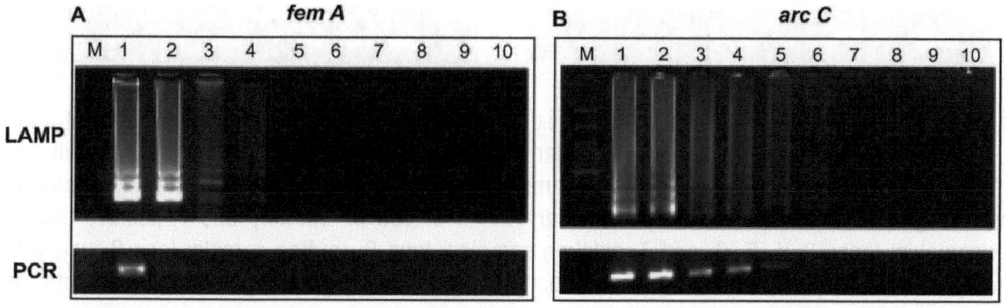

Fig. 2 Sensitivity of LAMP assay and PCR for *femA* and *arcC* genes with tenfold dilutions of DNA. (**a**) for *femA* and (**b**) for *arcC* genes. Lanes M, molecular marker of 100 bp marker; lane 1, 10^{-1}; lane 2, 10^{-2}; lane 3, 10^{-3}; lane 4, 10^{-4}; lane 5, 10^{-5}; lane 6, 10^{-6}; lane 7, 10^{-7}; lane 8, 10^{-8}; lane 9, 10^{-9}; lane 10, 10^{-10}. All the products were electrophoresed on 2.0% agarose gel and stained with ethidium bromide

forward primer and reverse primer, 1.0 μL DNA and 5.0 μL distilled water to make up the volume to 12.5 mL.

4. Set the PCR cycles to 35, with annealing at 57 °C for *femA* and 58 °C for *arcC* for 1 min.
5. For visualization, run gel electrophoresis using a 2% agarose gel and visualize the results using a Gel Doc system.

3.4 Screening of Staphylococcus aureus from Food Samples

1. Collect food samples (milk and dairy products, meat samples) in sterile 2.0 mL microcentrifuge tubes.
2. Homogenize 1 g or 1 mL of the sample in 1 mL of Butterfield's phosphate-buffered dilution water.
3. Spread 100 μL of the homogenized sample onto Baird-Parker agar (BPA) plates and incubate at 37 °C for 48 h.
4. After incubation, observe for the formation of black-centered colonies with clear zones. Streak these colonies into Luria-Bertani (LB) agar plates.
5. To confirm the isolates, perform a Latex agglutination assay, which screens for protein A and bound coagulase, a known marker for *S. aureus*.
6. Enrich positive samples identified by the conventional method by inoculating 1 g of sample into 9 mL of tryptic soy broth.
7. Incubate the samples for 12–18 h. Following incubation, subject the enrichment to crude DNA extraction by treating the

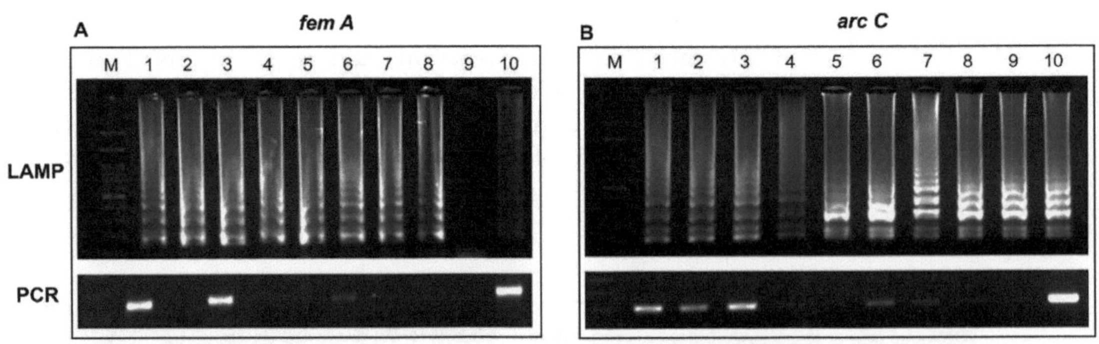

Fig. 3 Representative image of a LAMP and PCR screening of food samples for (**a**) *femA* and (**b**) *arcC* gene. LAMP screening of the samples, lane M, molecular marker of 100 bp marker; lanes 1 and 2, milk samples; lanes 3, 4, 5, 6, and 7, chicken samples; lane 8, mutton sample; lane 9, egg yolk; lane 10, bakery product. PCR screening for the samples, lane M, molecular marker of 100 bp marker; lane 1, positive control; lane 2, milk samples; lanes 3, 4, 5, 6, and 7, chicken samples; lane 8, mutton sample; lane 9, egg yolk; lane 10, bakery product. All the products were electrophoresed on 2.0% agarose gel and stained with ethidium bromide

sample with 1% Triton-100 and boiling for 10 min [8]. Use the supernatant obtained from this extraction as a template for both LAMP and PCR reactions.

8. For visualization, run gel electrophoresis using a 2% agarose gel and visualize the results using a Gel Doc system (Fig. 3).

4 Notes

1. Use gloves and a flame while handling bacterial cultures and discard the bacterial waste in designated culture discard bottles only.
2. Autoclave the Luria-Bertani broth before the experiment to sterilize the media.
3. Ensure primers do not overlap to prevent cross-amplification and false positives.
4. LAMP assay should be performed in sterile conditions to avoid contaminations.

Acknowledgments

This work was supported by the Department of Science and Technology (DST), Government of India under the Women Scientist Scheme (WOS-A): SR/WOS-A/LS-1065/2014 (G). The corresponding author also sincerely acknowledge Prof. Dr. Indrani Karunasagar for constant guidance throughout the project.

References

1. Alarcon B, Vicedo B, Aznar R (2006) PCR-based procedures for detection and quantification of *Staphylococcus aureus* and their application in food. J Appl Microbiol 2:352–364
2. Su J, Liu X, Cu H, Li Y, Chen D, Li Y, Yu G (2014) Rapid and simple detection of methicillin-resistance *Staphylococcus aureus* by orfX Loop-mediated isothermal amplification assay. BMC Biotechnol 14:8
3. Rodríguez-Lázaro D, Oniciuc EA, García PG, Gallego D, Fernández-Natal I, Dominguez-Gil M, Eiros-Bouza JM, Wagner M, Nicolau AI, Hernández M (2017) Detection and characterization of *Staphylococcus aureus* and methicillin-resistant *S. aureus* in foods confiscated in EU borders. Front Microbiol 8:1344
4. Goto M, Hayashidani H, Takatori K, Hara-Kudo Y (2007) Rapid detection of enterotoxigenic *Staphylococcus aureus* harbouring genes for four classical enterotoxins, SEA, SEB, SEC and SED, by loop-mediated isothermal amplification assay. Lett Appl Microbiol 45:100–107
5. Byrne B, Gilmartin N, Lakshmanan RS, O'Kennedy R (2015) Antibodies, enzymes, and nucleic acid sensors for high throughput screening of microbes and toxins in food. In: High throughput screening for food safety assessment, pp 25–80
6. Tomita N, Mori Y, Kanda H, Notomi T (2008) Loop-mediated isothermal amplification (LAMP) of gene sequences and simple visual detection of products. Nat Protoc 3:877–882
7. Ausubel FM, Kingston R, Moore RE, Seidman DD, Smith, J G JA, Struhl K, Ferreira

Behravesh CB, Williams, IT, Tauxe RV (2012) Emerging foodborne pathogens and problems: expanding prevention efforts before slaughter or harvest. Improving Food Safety through a One Health Approach Workshop Summary

8. Sowmya N, Thakur MS, Manonmani HK (2012) Rapid and simple DNA extraction method for the detection of enterotoxigenic *Staphylococcus aureus* directly from food samples: comparison of PCR and LAMP methods. J Appl Microbiol 113:106–113. https://doi.org/10.1111/j.1365-2672.2012.05315.x
9. Omoe K, Ishikawa M, Shimoda Y, Hu DL, Ueda S, Shinagawa K (2002) Detection of seg, *seh*, and *sei* genes in *Staphylococcus aureus* isolates and determination of the enterotoxin productivities of *S. aureus* isolates harbouring *seg*, *seh*, or *sei* genes. J Clin Microbiol 40:857–862
10. Zhang T, Wang C, Wei X, Zhao X, Zhong X (2012) Loop-mediated isothermal amplification for detection of *Staphylococcus aureus* in dairy cow suffering from mastitis. J Biomed Biotechnol:435982
11. Koide Y, Maeda H, Yamabe K, Naruishi K, Yamamoto T, Kokeguchi S, Takashiba S (2010) Rapid detection of *mecA* and *spa* by the Loop-mediated isothermal amplification (LAMP) method. Lett Appl Microbiol 50(4): 386–392. https://doi.org/10.1111/j.1472-765X.2010.02806.x. Epub 2010 Jan 22
12. Chen C, Zhao Q, Guo J, Li Y, Chen Q (2017) Identification of methicillin-resistant *Staphylococcus aureus* (MRSA) using simultaneous detection of *mecA*, *nuc*, and femB by Loop-mediated isothermal amplification (LAMP). Curr Microbiol 74(8):965–971. https://doi.org/10.1007/s00284-017-1274-2. Epub 2017 Jun 2

Chapter 8

Loop-Mediated Isothermal Amplification for the Detection of *Cronobacter sakazakii* in Milk Powder

Md Iqbal Hossain, Daseul Yeo, Hyojin Kwon, Dong Joo Seo, and Changsun Choi

Abstract

Loop-mediated isothermal amplification (LAMP) for the detection of *Cronobacter sakazakii* in milk powder was successfully developed. For the reliable detection of *C. sakazakii*, LAMP primer sets targeting the outer membrane protein A (*ompA*) gene of *C. sakazakii* were designed using Explorer V4 software. Thirty-seven *C. sakazakii* strains and 13 pathogenic microorganisms were used for comparative detection of *C. sakazakii* via polymerase chain reaction (PCR), real-time PCR, and LAMP. The LAMP method developed in this study could specifically detect *C. sakazakii* strains without cross-reactivity with other foodborne pathogens. LAMP products amplified from the *ompA* gene of *C. sakazakii* were digested with *Hha*I and *Nru*I enzymes. The specificity of LAMP was confirmed by restriction fragment length polymorphism (RFLP) analysis. LAMP could detect *C. sakazakii* within 1 h without the need for bacterial culture, with a detection limit as low as 1 CFU/mL of *C. sakazakii* in milk. In terms of sensitivity, LAMP showed a detection limit 10,000 times and 100 times lower than that of PCR and real-time PCR, respectively. Therefore, this study concludes that LAMP is a rapid and reliable detection technique for identifying *C. sakazakii* contamination in powdered milk.

Keywords Loop-mediated isothermal amplification (LAMP), *Cronobacter sakazakii*, *Outer membrane protein A (ompA) gene*, Specificity, Sensitivity

1 Introduction

Cronobacter sakazakii is a member of the Enterobacteriaceae family; it is a Gram-negative bacillus and is a facultative anaerobe that does not form spores [1]. *C. sakazakii* can cause such as meningitis, sepsis, and neonatal necrotizing enterocolitis in all age groups, especially newborns under 1-month-old, premature babies, low birth weight babies, and young children. It is a very dangerous acute opportunistic infection that causes symptoms [2, 3].

Existing detection methods for *C. sakazakii* include bacterial culture, biochemical testing, and serological analysis, but these methods have the problem of being time-consuming and

Alejandro Garrido-Maestu and Alexandre Lamas (eds.), *Foodborne Pathogens: Loop-mediated Isothermal Amplification Protocols*, Methods and Protocols in Food Science, https://doi.org/10.1007/978-1-0716-4526-0_8,

inaccurate [4]. The US Food and Drug Administration (US FDA) recommended bacterial isolation and culture, polymerase chain reaction (PCR), and real-time PCR methods as detection methods for *C. sakazakii*. Compared to bacterial isolation and culture methods, which are time-consuming and have low sensitivity, PCR has good sensitivity and specificity, but has the disadvantage of being difficult to use universally because it requires an electrophoresis process and a professional technician to confirm DNA [5]. Real-time PCR has higher sensitivity and DNA amplification accuracy than PCR and can be confirmed quickly and accurately using fluorescent substances without electrophoresis process, but the disadvantage is that it is sophisticated and complex and requires expensive equipment [6, 7].

Recently, a loop-mediated isothermal amplification (LAMP) method was developed that can sensitively and rapidly amplify DNA under isothermal condition [8]. The LAMP detection method is similar to PCR in that it amplifies the target gene, but can shorten the reaction time because it allows amplification of a specific gene at a constant temperature without temperature change. LAMP can amplify a large amount of target genes using only four or six primers specifically designed based on *Bst* DNA polymerase and the specific base sequence of the target DNA and some simple equipment under isothermal conditions for denaturation of DNA strands [8]. This LAMP technique is being used as a rapid detection technology for pathogenic microorganisms such as *Campylobacter jejuni*, *Campylobacter coli*, *Mycobacterium avium*, *Yersinia pseudotuberculosis*, *Mycoplasma pneumoniae*, and *Porphyromonas gingivalis* and can be used for early detection and prevention of food poisoning pathogens [7, 9–11].

Therefore, this study developed LAMP technology that can specifically amplify *C. sakazakii* in milk powder.

2 Materials

2.1 Bacterial Strains

1. Two *Cronobacter sakazakii* standard strains: *C. sakazakii* ATCC 29004 and *C. sakazakii* ATCC 29544 (Table 1).
2. Thirty-five *C. sakazakii* strains: 35 *C. sakazakii* strains were obtained from the Department of Food and Nutrition at Chung-Ang University Provided by the Food Safety and Virology Laboratory (Table 1).
3. Negative control strains (13 pathogenic microorganisms): *Arcobacter buzleri* ATCC 49616, *Bacillus cereus* ATCC 13061, *B. cereus* ATCC 10876, *Escherichia coli* O157:H7 ATCC 43890, *E. coli* O157:H7 ATCC 43889, *Listeria monocytogenes* ATCC 19114, *L. monocytogenes* ATCC 15313, *Salmonella* Typhimurium ATCC 43174, *Salmonella* Enteritidis

Table 1
Comparison of PCR and LAMP detection in the reference bacterial strains used in the study

No.	Original ID	Source Institution	LAMP	PCR	No.	Original ID	Source Institution	LAMP	PCR
1	*Cronobacter sakazakii* 1	FDA	+	+	26	*C. sakazakii* FSM 292	FDA	+	+
2	*C. sakazakii* 2	FDA	+	+	27	*C. sakazakii* FSM 293	FDA	+	+
3	*C. sakazakii* 3	FDA	+	+	28	*C. sakazakii* FSM 294	FDA	+	+
4	*C. sakazakii* 5	FDA	+	+	29	*C. sakazakii* FSM 295	FDA	+	+
5	*C. sakazakii* 6	FDA	+	+	30	*C. sakazakii* FSM 297	FDA	+	+
6	*C. sakazakii* 2.40	FDA	+	+	31	*C. sakazakii* FSM 298	FDA	+	+
7	*C. sakazakii* 2.44	FDA	+	+	32	*C. sakazakii* FSM 299	FDA	+	+
8	*C. sakazakii* 2.46	FDA	+	+	33	*C. sakazakii* FSM 300	FDA	+	+
9	*C. sakazakii* 2.47	FDA	+	+	34	*C. sakazakii* FSM 302	FDA	+	+
10	*C. sakazakii* 2.80	FDA	+	+	35	*C. sakazakii* FSM 303	FDA	+	+
11	*C. sakazakii* 2.81	FDA	+	+	36	*C. sakazakii* 29004	ATCC	+	+
12	*C. sakazakii* 2.82	FDA	+	+	37	*C. sakazakii* 29544	ATCC	+	+
13	*C. sakazakii* 2.83	FDA	+	+	38	*Arcobacter butzleri* 49616	ATCC	–	–
14	*C. sakazakii* 2.84	FDA	+	+	39	*Bacillus cereus* 10876	ATCC	–	–
15	*C. sakazakii* 279–1	FDA	+	+	40	*B. cereus* 13061	ATCC	–	–
16	*C. sakazakii* FSM 262	FDA	+	+	41	*Escherichia coli* O157: H7 43889	ATCC	–	–
17	*C. sakazakii* FSM 265	FDA	+	+	42	*E. coli* O157:H7 43890	ATCC	–	–
18	C. sakazakii FSM 270	FDA	+	+	43	*Listeria monocytogenes* 15313	ATCC	–	–
19	*C. sakazakii* FSM 271	FDA	+	+	44	*L. monocytogenes* 19114	ATCC	–	–
20	*C. sakazakii* FSM 272	FDA	+	+	45	*Salmonella* Enteritidis 13076	ATCC	–	–

(continued)

Table 1
(continued)

No.	Original ID	Source Institution	LAMP	PCR	No.	Original ID	Source Institution	LAMP	PCR
21	*C. sakazakii* FSM 273	FDA	+	+	46	*Salmonella* Typhimurium 43971	ATCC	–	–
22	*C. sakazakii* FSM 274	FDA	+	+	47	*Staphylococcus aureus* 10768	KCCM	–	–
23	*C. sakazakii* FSM 275	FDA	+	+	48	*S. aureus* 12600	ATCC	–	–
24	*C. sakazakii* FSM 287	FDA	+	+	49	*Vibrio parahaemolyticus* 43996	ATCC	–	–
25	*C. sakazakii* FSM 290	FDA	+	+	50	*V. parahaemolyticus* 4664	KACC	–	–

This data was reproduced by the approval of a previous publication [12]

ATCC 43076, *Staphylococcus aureus* ATCC 12600, *S. aureus* ATCC 10768, *Vibrio parahaemolyticus* ATCC 4664, and *V. parahaemolyticus* ATCC 43996 from the American Type Culture Collection (ATCC; USA). Each strain (13 pathogenic microorganisms) was used as a negative control (Table 1).

2.2 Growth Media for Bacterial Strain

1. *C. sakazakii* strains were cultured using Tryptic Soy Broth (TSB; Difco, Becton Dickinson, USA) and Brilliance™ Enterobacter Sakazakii Agar (Oxoid, United Kingdom).
2. *A. buzleri* was cultured on Arcobacter Selective Broth (Oxoid) by adding 1.2% Agar Technical (Oxoid), *B. cereus* on Tryptic Soy Agar (Oxoid), *E. coli* O157:H7 on Sorbitol McConkey Agar (Oxoid), *L. monocytogenes* on Listeria Selective Agar Base (Oxoid), *Salmonella* spp. on Rappaport-Vassiliadis Soya Peptone Agar (Oxoid), *S. aureus* on Baird-Parker Agar (Oxoid), and *V. parahaemolyticus* on Tryptic Soy Agar (TSA) containing 2% NaCl.

2.3 DNA Extraction

1. 1.5 mL Eppendorf tube.
2. Water bath (Daihan Scientific Co., Wonju, Korea).
3. Ice.
4. Centrifuge machine (Allegra™ 64 R Benchtop Centrifuge, Beckman-Coulter, Brea, CA).

Table 2
Oligonucleotide sequences of PCR, real-time PCR, and LAMP for the detection of *C. sakazakii* outer membrane protein A (*ompA*) gene

	Primer	Oligonucleotide sequence	References
	F3	5′-CCC GGT GAA GGA TTT AAC CG-3′	
LAMP	B3	5′-CGT CGT TAG GGA TAA AGC CG-3′	[12]
	BIP	5′-CAG GCC GCA CCG AAA GAT AAC AAT CGT GGA ACT GGG ACC A-3′	
	FIP	5′-TGC AAT CGC GAT AGC CGT CTT TCA AGG AAA AGC GCA TGG C-3′	
	Forward	5′-CCG GAA CAA GCT GAA AAT TGA-3′	
Real-time PCR	Reverse	5′-TCT TCG TGC TGC GAG TTT G-3′	[6]
	Probe	FAM-5′-ACT CTG ACA CAC CGC GCA TTC CTG-3′-TAMRA	

2.4 Development of LAMP Detection Method for *C. sakazakii*

1. Outer membrane protein A (*ompA*) gene (Genbank accession number: GQ845410).
2. Online LAMP primer design software PrimerExplorer V4 (https://primerexplorer.eiken.co.jp/e/).
3. 1× ThermoPol reaction buffer (New England BioLabs Inc., Ipswich, United Kingdom) (*see* **Note 1**)
4. Deoxynucleoside triphosphate (dNTP; Bioneer, Daejeon, Republic of Korea).
5. 8 U of *Bst* DNA polymerase (New England BioLabs Inc.)
6. Inner primers FIP and BIP (Table 2).
7. Outer primers F3 and B3 (Table 2).
8. Template DNA.
9. Diethylpyrocarbonate (DEPC)-treated deionized water.
10. MJ Mini Personal Thermal Cycler (Bio-Rad, USA).
11. 1% agarose gel using polyacrylamide gel electrophoresis (Bioneer)
12. Launch DocIT LS (UVP Advanced BioImaging Software version. 5.5.5a, UK).

2.5 Specificity Test of LAMP Detection Method

1. GeneCutter (DNASTAR).
2. Restriction fragment length polymorphism (RFLP) analysis.
3. Restriction enzymes *Hha*I and *Nru*I.
4. 1× ThermoPol reaction buffer
5. 10 U of *Nru*I or *Hha*I
6. Bovine serum albumin (BSA).

7. LAMP product DNA.
8. DEPC-treated deionized water.
9. Launch DocIT LS (UVP Advanced BioImaging Software version. 5.5.5a, UK).

2.6 To Verify the Specificity of the C. sakazakii LAMP Detection Method

1. Positive strains: 37 of *C. sakazakii* strains, including standard strains.
2. Negative controls: 13 pathogenic microorganisms.
3. The LAMP detection method used was the *C. sakazakii* LAMP detection method developed in this study, and PCR was performed as follows:
4. For PCR:
   (a) 10× reaction buffer (Bioneer, Daejeon, Korea) (*see* **Note 2**).
   (b) dNTP.
   (c) *Top* polymerase (Bioneer).
   (d) Template DNA.
   (e) The designed outer primers F3 and B3 were used as PCR primers to develop the LAMP detection method (Table 2).
   (f) 1% agarose gel electrophoresis (Bioneer).
   (g) Launch DocIT LS BioImaging Software.

2.7 Comparison of Sensitivity for Detection of C. sakazakii *in Powder Milk Between LAMP, PCR, and Real-Time PCR*

1. Actual powder milk or powder-type foods, commercially available powder milk was purchased.
2. *C. sakazakii* ATCC 29004 standard strain.
3. Brain Heart Infusion Broth (Sigma-Aldrich).
4. The powder milk was made 20% with 0.1% bacteriological peptone water (PW).
5. Genomic DNA Extraction kit (Bioneer).
6. Real-time PCR: Premix Ex *Taq*™ (2×) buffer (Takara Biotech, China), forward and reverse primers (Table 2) that amplify the *C. sakazakii* 16S–23S internal transcribed spacer (ITS) as the target gene, TaqMan probe (Table 2), template DNA, DEPC-treated deionized water.
7. Thermal Cycler Dice Real-Time System (Takara, Japan).

3 Methods

Carry out all procedures at room temperature unless otherwise specified.

3.1 Bacterial Strains in Growth Media

1. Primary culture of *C. sakazakii* strain was performed on Brilliance™ Enterobacter Sakazakii Agar for 24 h at 37 °C.
3. Primary culture of negative control strains: *A. buzleri* on Arcobacter Selective Broth by adding 1.2% Agar Technical, *B. cereus* on TSA, *E. coli* O157:H7 on Sorbitol McConkey Agar, *L. monocytogenes* on Listeria Selective Agar Base, *Salmonella* spp. on Rappaport-Vassiliadis Soya Peptone Agar, *S. aureus* on Baird-Parker Agar, and *V. parahaemolyticus* on TSA containing 2% NaCl.
4. A single colony was selected through primary culture on specified agar medium and was cultured secondary in TSB for 24 h at 37 °C and used for DNA extraction.

3.2 DNA Extraction

1. After secondary culture, 37 *C. sakazakii* and 13 negative control strains culture (1 mL) were each transferred to a 1.5 mL Eppendorf tube.
2. The cultures were centrifuged at 13000 rpm for 15 min.
3. The supernatant was discarded, and the bacterial pellet was resuspended in 100 μL of sterile water.
4. Boiled in water bath at 100 °C for 10 min to lyse the cells, and then cooled on ice.
5. The suspension was centrifuged at 8000 rpm for 2 min, and the supernatant was transferred to a new 1.5 mL Eppendorf tube and the supernatant was used as a DNA template of each strain.

3.3 Development of LAMP Detection Method for C. sakazakii

1. After securing a large number of primer candidates capable of amplifying the *C. sakazakii ompA* gene, the optimal primer (Table 2) without nonspecific reaction was selected in preliminary research. Therefore, it was used in this study.
2. To perform the LAMP detection method: 2.5 μL of 1× ThermoPol reaction buffer, 2 μL of 10 mM dNTPs mixture, 1 μL of *Bst* DNA polymerase (8 U), 2 μL of FIP primer (40 pmol), 2 μL of BIP primer (40 pmol), 1 μL of F3 primer (20 pmol), 1 μL of B3 primer (20 pmol), and 2 μL of template DNA, and the final reaction volume was adjusted to 25 μL with 11.5 μL DEPC-treated distilled water.
3. The reaction mixture was incubated at 64 °C for 60 min and at 80 °C for 10 min using an MJ Mini Personal Thermal Cycler for temperature control.
4. LAMP products were visualized on a 1% agarose gel using polyacrylamide gel electrophoresis and analyze on Launch DocIT LS BioImaging Software (see the result on Table 1).

3.4 Specificity Test of LAMP Detection Method

1. Unlike the DNA products obtained by PCR, LAMP detection method produces a continuous DNA amplification product, so it appears in a ladder-like type when electrophoresis was performed.
2. To test the specificity of this DNA amplification product, the specific base sequence present in the amplification site was confirmed using GeneCutter (DNASTAR).
3. The region identified in the base sequence analysis of the *C. sakazakii ompA* amplification product was tested for the specificity by performing RFLP using *Hha*I and *Nru*I restriction enzymes.
4. For RFLP: 1× ThermoPol reaction buffer, 10 U *Nru*I or *Hha*I, 2 μg of BSA, 4 μL of LAMP product DNA. After forming the reaction mixture with LAMP product DNA, the final reaction volume was adjusted to 20 μL using DEPC-treated distilled water.
5. RFLP was reacted at 37 °C for 2 h and at 65 °C for 25 min to inactivate the enzyme.
6. The results were analyzed using Launch DocIT LS BioImaging Software after electrophoresis (see the result in Fig. 1).

3.5 To Verify the Specificity Test of the C. sakazakii LAMP Detection Method

1. To verify the specificity of the *C. sakazakii* LAMP detection method for positive and negative controls, LAMP and PCR were performed on 37 *C. sakazakii* and 13 pathogenic microorganisms, including standard strains (see the results on Table 1 and Fig. 2).
2. The LAMP detection method used was the *C. sakazakii* LAMP detection method developed in this study, and PCR was performed as follows.
3. PCR mixture (per reaction): 2 μL of 10× reaction buffer, 2 μL of 10 mM concentration of dNTPs, 1 μL of *Top* polymerase (1 U), 2 μL template DNA, 2 μL of each forward (20 pmol) and reverse primer (20 pmol) (outer primers F3 and B3 were used as PCR primers to develop the LAMP detection method (Table 2)), and 11 μL with DEPC-deionized water added to bring the total reaction volume to 20 μL.
4. The PCR was performed by initial denaturing the DNA at 94 ° C for 5 min (1 cycle), followed by 35 cycles of denaturation at 94 °C for 30 s, annealing at 57 °C for 30 s, extension at 72 °C for 30 s, and final extension at 72 °C for 5 min (1 cycle).
5. PCR products were visualized on a 1% agarose gel electrophoresis and analyzed using Launch DocIT LS BioImaging Software (see the result in Fig. 2).

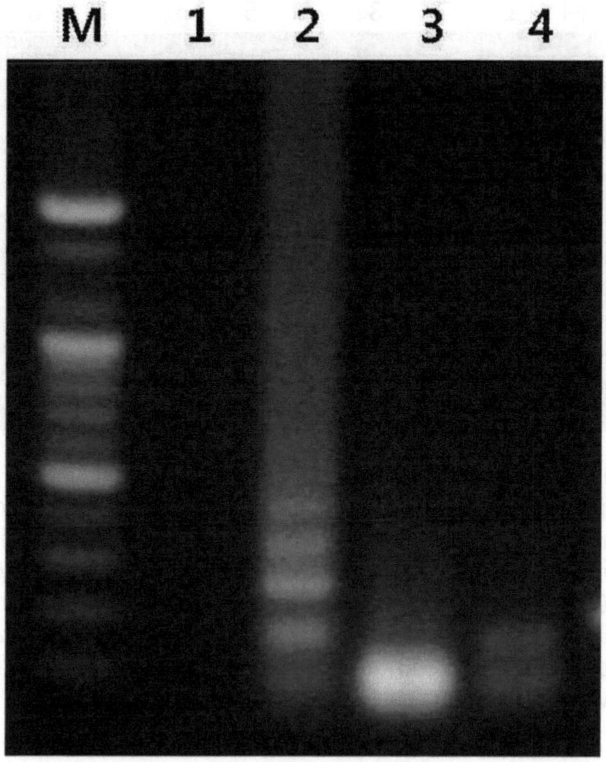

Fig. 1 Restriction fragment length polymorphism (RFLP) of LAMP DNA products using *Hha*I and *Nru*I restriction enzyme. M: 100-bp DNA ladder, Lane 1: negative control, Lane 2: *C. sakazakii* LAMP product. Lane 3: digestion of *C. sakazakii* LAMP product with *Hha*I restriction enzyme, Lane 4: digestion of *C. sakazakii* LAMP product with *Nru*I restriction enzyme. (This data was reproduced by the approval of the previous publication [12])

3.6 Comparison of Sensitivity for Detection of C. sakazakii in Powder Milk Between LAMP, PCR, and Real-Time PCR

1. To confirm the possibility of detecting *C. sakazakii* bacteria in actual powder milk, a detection sensitivity experiment was conducted.
2. The inoculum was *C. sakazakii* ATCC 29004, a standard strain, cultured in Brain Heart Infusion Broth.
3. After culturing for 24 h at 37 °C, the number of bacteria was measured and an inoculum was prepared.
4. Powder milk (20%) was made with 0.1% bacteriological peptone water (PW).
5. *C. sakazakii* ATCC 29004 was added to 10 mL of the powder milk at 7 Log, 6 Log, 5 Log, 4 Log, 3 Log, 2 Log, 1 Log, 0 Log, and −1 Log CFU/mL level of artificial contamination (see the result in Fig. 3).
6. DNA from samples artificially inoculated with *C. sakazakii* was extracted using a Genomic DNA Extraction kit.

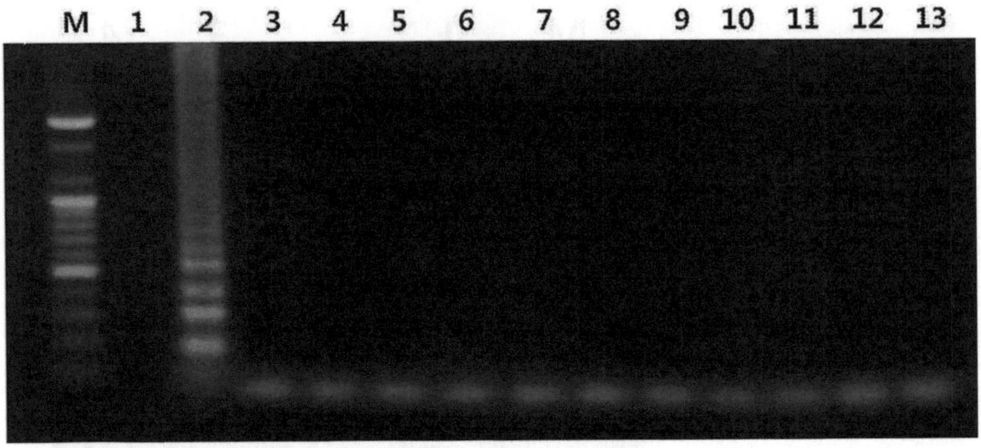

Fig. 2 Specificity of loop-mediated isothermal amplification (LAMP) for the detection of *Cronobacter sakazakii* using reference strains. M: 100-bp DNA ladder, Lane 1: negative control, Lane 2: *Cronobacter sakazakii* ATCC 29004, Lane 3: *Salmonella* Enteritidis ATCC 13076, Lane 4: *Salmonella* Typhimurium ATCC 43971, Lane 5: *Arcobacter butzleri* ATCC 49616, Lane 6: *Listeria monocytogenes* ATCC 19114, Lane 7: *Listeria monocytogenes* ATCC 15313, Lane 8: *Vibrio parahaemolyticus* KCCM 4664, Lane 9: *Vibrio parahaemolyticus* ATCC 43996, Lane 10: *Escherichia coli* O157:H7 ATCC 43889, Lane 11: *Escherichia coli* O157:H7 ATCC 43890, Lane 12: *Bacillus cereus* ATCC 13061, Lane 13: *Staphylococcus aureus* ATCC 12600. (This data was reproduced by the approval of the previous publication [12])

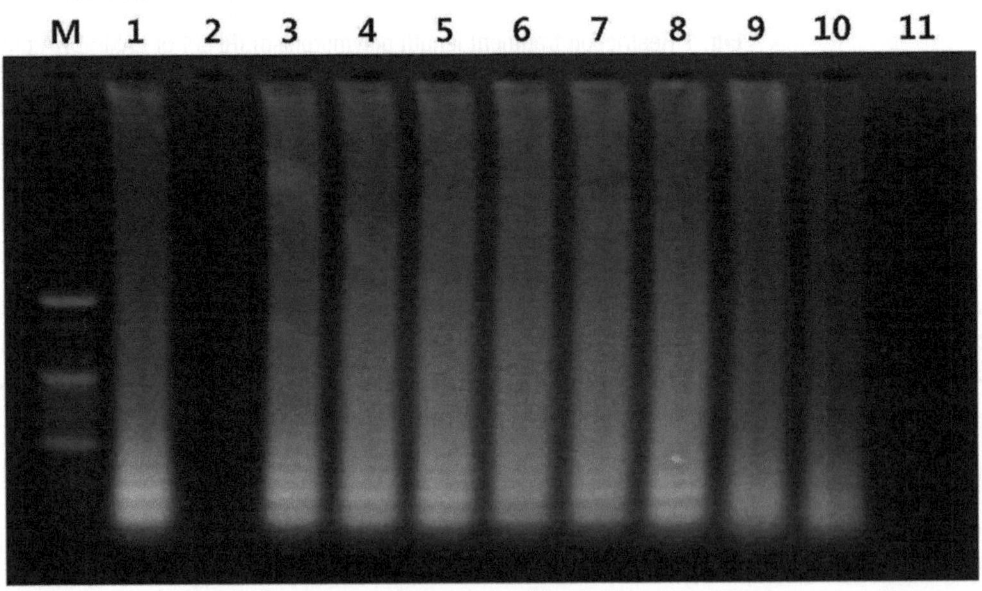

Fig. 3 Sensitivity of loop-mediated isothermal amplification (LAMP) for the detection of *C. sakazakii* in milk powder. M: 100-bp DNA ladder, Lane 1: positive control, Lane 2: negative control, Lane 3: 7 Log CFU/mL *C. sakazakii* in milk powder, Lane 4: 6 Log CFU/mL *C. sakazakii* in milk powder, Lane 5: 5 Log CFU/mL *C. sakazakii* in milk powder, Lane 6: 4 Log CFU/mL *C. sakazakii* in milk powder, Lane 7: 3 Log CFU/mL *C. sakazakii* in milk powder, Lane 8: 2 Log CFU/mL *C. sakazakii* in milk powder, Lane 9: 1 Log CFU/mL *C. sakazakii* in milk powder, Lane 10: 0 Log CFU/mL *C. sakazakii* in milk powder, Lane 11: −1 Log CFU/mL *C. sakazakii* in milk powder. (This data was reproduced by the approval of the previous publication [12])

7. LAMP, PCR, and real-time PCR were performed respectively to compare and analyze the detection sensitivity of *C. sakazakii.*
8. The LAMP and PCR methods were performed in the same manner as the specificity test for *C. sakazakii*, and the detection sensitivity of real-time PCR was compared using the TaqMan probe method reported in a previous study by Liu et al. [6].
9. TaqMan-based real-time PCR reactions were performed in a 25-μL reaction volume containing (per reaction):
   (a) Premix Ex *Taq*™ (2×) buffer 12.5 μL.
   (b) Forward Primer (10 μM) 1 μL.
   (c) Reverse Primer (10 μM) 1 μL.
   (d) TaqMan probe (5 μM) 1 μL.
   (e) Template DNA 2 μL.
   (f) DEPC-treated deionized water to bring the final reaction volume to 25 μL.
10. The TaqMan-based real-time PCR were run on a Thermal Cycler Dice Real Time System (Takara) using the following program: denaturation at 95 °C for 10 s, 45 cycles consisted of denaturation at 95 °C for 10 s, annealing at 55 °C for 20 s, and extension at 72 °C for 15 s [6].

4 Notes

1. 1× ThermoPol reaction buffer: 20 mM Tris-HCl (pH 8.8), 10 mM $(NH_4)_2SO_4$, 10 mM KCl, 2 mM MgSO4, 0.1% Triton ×-100 (New England BioLabs Inc., Ipswich, UK)
2. 10× reaction buffer: Tris [pH 9.0], 15 mM $MgCl_2$ (Bioneer).

Acknowledgments

This research was supported by the National Research Foundation of Korea [NRF2018R1A6A1A03025159].

References

1. Friedemann M (2007) *Enterobacter sakazakii* in food and beverages (other than infant formula and milk powder). Int J Food Microbiol 116(1):1–10. https://doi.org/10.1016/j.ijfoodmicro.2006.12.018
2. Arad I, Baras M, Gofin R et al (2001) Does parity affect the neonatal outcome of very low birth weight infants. Eur J Obstet Gynecol Reprod Biol 94(2):283–288. https://doi.org/10.1016/S0301-2115(00)00308-0
3. Guillaume-Gentil O, Sonnard V, Kandhai MC et al (2005) A Simple and rapid cultural method for detection of *Enterobacter sakazakii* in environmental samples. J Food Prot 68(1):64–69. https://doi.org/10.4315/0362-028X-68.1.64

4. Derzelle S, Dilasser F, Maladen V et al (2007) Comparison of three chromogenic media and evaluation of two molecular-based identification systems for the detection of *Enterobacter sakazakii* from environmental samples from infant formulae factories. J Food Prot 70(7): 1678–1684. https://doi.org/10.4315/0362-028X-70.7.1678
5. Orlandi PA, Lampel KA (2000) Extraction-free, filter-based template preparation for rapid and sensitive PCR detection of pathogenic parasitic protozoa. J Clin Microbiol 38(6):2271–2277. https://doi.org/10.1128/jcm.38.6.2271-2277.2000
6. Liu Y, Cai X, Zhang X et al (2006) Real time PCR using TaqMan and SYBR Green for detection of *Enterobacter sakazakii* in infant formula. J Microbiol Meth 65(1):21–31. https://doi.org/10.1016/j.mimet.2005.06.007
7. Yamazaki W, Taguchi M, Ishibashi M et al (2008) Development and evaluation of a Loop-mediated isothermal amplification assay for rapid and simple detection of *Campylobacter jejuni* and *Campylobacter coli*. J Med Microbiol 57(4):444–451. https://doi.org/10.1099/jmm.0.47688-0
8. Notomi T, Okayama H, Masubuchi H et al (2000) Loop-mediated isothermal amplification of DNA. Nucleic Acids Res 28(12):63e. https://doi.org/10.1093/nar/28.12.e63
9. Enosawa M, Kageyama S, Sawai K et al (2003) Use of Loop-mediated isothermal amplification of the IS900 sequence for rapid detection of cultured *Mycobacterium avium* subsp. paratuberculosis. J Clin Microbiol 41(9): 4359–4365. https://doi.org/10.1128/jcm.41.9.4359-4365.2003
10. Saito R, Misawa Y, Moriya K et al (2005) Development and evaluation of a Loop-mediated isothermal amplification assay for rapid detection of *Mycoplasma pneumoniae*. J Med Microbiol 54(11):1037–1041. https://doi.org/10.1099/jmm.0.46071-0
11. Yoshida A, Nagashima S, Ansai T et al (2005) Loop-mediated isothermal amplification method for rapid detection of the periodontopathic bacteria *Porphyromonas gingivalis*, *Tannerella forsythia*, and *Treponema denticola*. J Clin Microbiol 43(5):2418–2424. https://doi.org/10.1128/jcm.43.5.2418-2424.2005
12. Kim Y-J, Seo S, Wang X, Seo DJ, Lee MH, Son NR, Lee B-H, Choi C (2013) Comparison of polymerase chain reaction, real-time Polymerase Chain Reaction, and Loop-mediated isothermal amplification for the detection of *Cronobacter sakazakii* in milk powder. Korean J Food Sci An 33(5):610–616. https://doi.org/10.5851/kosfa.2013.33.5.610

Chapter 9

Rapid Detection of *Salmonella* ser. Enteritidis by Loop-Mediated Isothermal Amplification (LAMP) Method

Lijun Hu and Guodong Zhang

Abstract

Salmonella enterica serovar Enteritidis represents a significant global public health concern. It is recognized as the most prevalent *Salmonella* serotype responsible for salmonellosis outbreaks associated with poultry, eggs, and egg-related products. The loop-mediated isothermal amplification (LAMP) technique is an innovative and user-friendly detection method that rapidly, efficiently, and specifically amplifies DNA under constant temperature conditions. Here we present a LAMP protocol designed for the detection of *Salmonella* ser. Enteritidis using egg products as an example, which could be verified and extended to other types of samples. This protocol has been validated internally for inclusivity and exclusivity, with a detection limit of 1 CFU/25 g egg products and 1 CFU/1000 g liquid eggs with enrichment. The run time is about 35 min. This LAMP method allows the simultaneous analysis of up to eight samples including controls in a single run.

Key words LAMP, *Salmonella* Enteritidis, Egg, Detection, Protocol

1 Introduction

Since the advent of the loop-mediated isothermal amplification (LAMP) technology in 2000 [1], it has gained widespread use and become a significant tool for testing pathogens in clinical, animal, food, and environmental samples. LAMP facilitates the rapid amplification and detection of nucleic acids in a single-step process, involving two stages: noncyclic and cyclic steps (Fig. 1) [2, 3, 12]. The process begins with the creation of DNA possessing stem loops at both ends, acting as the initial structure for amplification during the noncyclic step. Subsequently, the dumbbell-like DNA structure undergoes rapid transformation into a stem-loop DNA through self-primed DNA synthesis. This iterative process allows for substantial amplification, with the potential to increase DNA copies up to 10^9–10^{10} times in 15–60 min. The optimal temperature for amplification occurs at 60–65 °C. The reaction

Alejandro Garrido-Maestu and Alexandre Lamas (eds.), *Foodborne Pathogens: Loop-mediated Isothermal Amplification Protocols*, Methods and Protocols in Food Science, https://doi.org/10.1007/978-1-0716-4526-0_9,

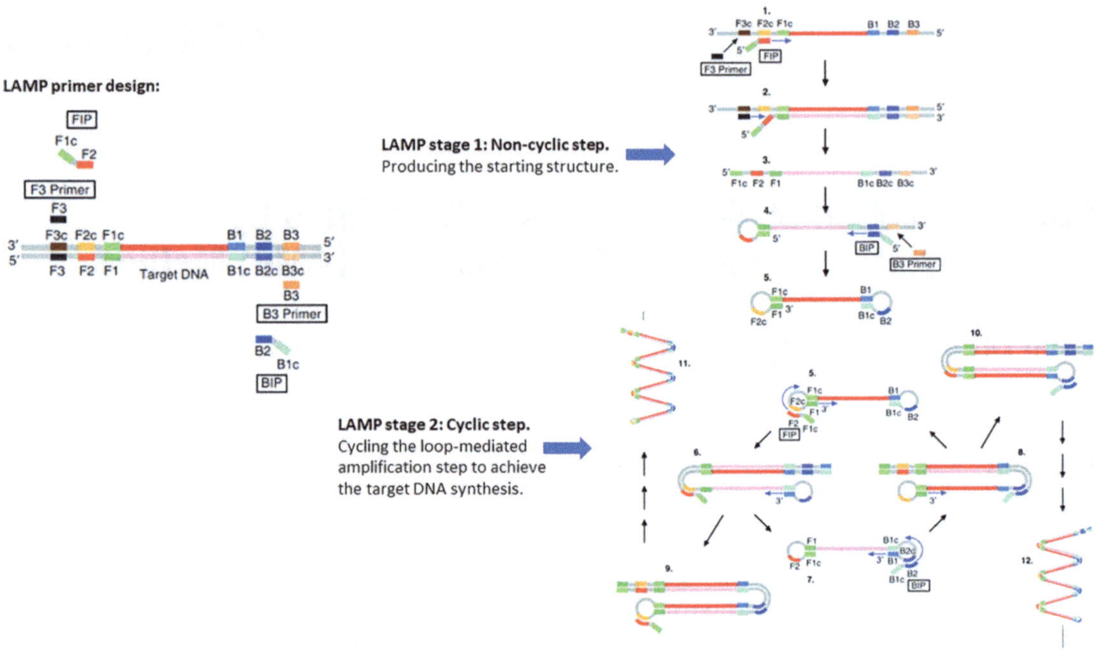

Fig. 1 Basic principle of LAMP method. (The figure was regenerated followed the previous publications [2, 3, 12])

mixture contains the amplification buffer, dNTPs, nucleic acid sample (DNA/RNA), specifically designed primers (4/6 primers), *Bst* DNA polymerase, or equivalent. A typical LAMP assay employs a primer set comprising two inner primers (FIP and BIP) and two outer primers (F3 and B3) that recognize six distinct regions of the target DNA sequence. Additionally, forward and backward (LF and LB) loop primers can be designed and added to the reaction tube to enhance the LAMP process by accelerating DNA synthesis from increased starting points [1, 4].

This distinctive nucleic acid amplification technology confers the assay with quick, user-friendly, and cost-effective attributes compared to other methods for foodborne pathogens detection, such as traditional culture method, thermal cycling-based methods (e.g., PCR and real-time PCR), metagenomics, etc. Additional advantages of LAMP are its high sensitivity and specificity, tolerance to PCR-inhibiting substances [5, 6], acceptance of broad pH and temperature ranges, capacity to handle non-processed samples, inexpensive devices needed (using only a heating block/water bath could conduct the performance), and flexibility in readout methods [7].

Salmonella enterica serovar Enteritidis (*Salmonella* ser. Enteritidis) represents a significant global public health concern. It is recognized as the most prevalent *Salmonella* serotype responsible for salmonellosis outbreaks associated with poultry, eggs, and egg-related products [8]. Based on the principles of LAMP technology, we have effectively developed a rapid and simple assay for detecting *Salmonella* ser. Enteritidis using the Genie®III portable device. The Genie III system is a closed-tube, touchscreen, real-time platform proficient in monitoring DNA amplification with two-color fluorescence excitations, which enables visual detection of amplification results, presenting real-time curves for eight samples per run. The incorporation of GPS, rechargeable battery, and Wi-Fi functionality into this portable device enhances the feasibility of conducting assays in field settings. Additionally, we used a new DNA polymerase known as GspSSD LF DNA polymerase, which significantly accelerated the LAMP reaction and facilitated the precise acquisition of detection results within 35 min [9]. Here, we present the step-by-step LAMP protocol for detecting *Salmonella* ser. Enteritidis using egg products as an example.

2 Materials

Ensure the workstation comprises a level table with ample surface area in a clean, well-lit, well-ventilated room, minimizing dust and drafts. Storage spaces must be dust and insect-free, with adequate capacity to safeguard equipment and supplies. Utilize molecular grade water (excluding autoclaved media) and analytical grade reagents for solution preparation. Follow precise procedures for reagent preparation at room temperature and store them according to specified temperature requirements. Adhere diligently to waste disposal regulations when discarding waste materials.

2.1 Equipment and Supplies

1. Genie® III (OptiGene Ltd., West Sussex, United Kingdom).
2. Genie 8-well strips (OptiGene Ltd., Catalog#: OP-0008).
3. Genie strip holder (OptiGene Ltd., Catalog#: GBLOCK).
4. Whirl-Pak® filter sterilized sampling bags (Whirl-Pak) or appropriate containers.
5. Microcentrifuge and microcentrifuge tubes (1.5 mL and 2.0 mL).
6. Vortex mixer.
7. Pipette and pipette tips (aerosol-resistant).
8. Analytical balance (accuracy, 0.001 g and 0.0001 g).
9. Heat block.
10. Sterile disposable inoculating needles and loops (1 μL and 10 μL), spreaders.
11. Sterile petri dishes (glass or plastic, size 15 × 100 mm).

12. Bacterial culture tubes (16 × 150 mm or 20 × 150 mm) and serological tubes (10 × 75 mm or 13 × 100 mm).
13. pH test paper and meter,
14. Incubator (35 ± 2 °C and 42 ± 0.2 °C).
15. Refrigerated incubator or laboratory refrigerator (4 ± 2 °C).
16. Culture tube racks and sampling bag racks.
17. Freezer (−20 °C or −80 °C).

2.2 Media and Reagents

1. Isothermal Mastermix (OptiGene Ltd., Catalog#: ISO-001).
2. InstaGene™ Matrix (BIO-RAD, Catalog#: 7326030).
3. DNA AWAY Surface Decontaminant (ThermoFisher Scientific, Catalog#: 7010-HIDPK) or equivalent.
4. UltraPure DNase/RNase-free distilled water (ThermoFisher Scientific, Catalog#: 10977015) or equivalent.
5. Trypticase (tryptic) soy broth: 17 g trypticase peptone, 3 g phytone peptone, 5 g NaCl, 2.5 g K_2HPO_4, 2.5 g glucose, 1 L distilled water. Heat with gentle agitation to dissolve. Autoclave 15 min at 121 °C. Final pH, 7.3 ± 0.2 (*see* **Note 1**).
6. Lactose broth: 3 g beef extract, 5 g peptone, 5 g lactose, 1 L distilled water. Autoclave 15 min at 121 °C. Final pH, 6.9 ± 0.2 (*see* **Note 2**).
7. Tetrathionate (TT) broth: (1) Tetrathionate broth base: 5 g polypeptone, 1 g bile salts, 10 g $CaCO_3$, 30 g $Na_2S_2O_3{\cdot}5H_2O$, 1 L distilled water. Suspend ingredients in 1 L distilled water, mix, and heat to boiling. Do not autoclave. Cool to less than 45 °C. Store at 5–8 °C. Final pH, 8.4 ± 0.2. (2) Iodine-potassium iodide (I_2-KI) solution: 5 g KI, 6 g I_2 (resublimed), 20 mL distilled water. Dissolve KI in 5 ml sterile distilled water. Add I_2 and stir to dissolve. Dilute to 20 mL. (3) Brilliant green solution: 0.1 g brilliant green dye (sterile), 100 ml distilled water.

   On day of use, add 20 mL I_2-KI solution and 10 mL brilliant green solution to 1 L TT broth base. Resuspend precipitate by gentle agitation. Aseptically dispense 10 mL portions into 20 × 150 or 16 × 150 mm sterile test tubes (*see* **Note 3**).
8. Rappaport-Vassiliadis (RV) broth: (1) Broth base: 5 g tryptone, 8 g NaCl, 1.6 g KH_2PO_4, 1 L distilled water. (2) Magnesium chloride ($MgCl_2$) solution: 400 g $MgCl_2{\cdot}6H_2O$, 1 L distilled water. (3) Malachite green oxalate solution: 0.4 g malachite green oxalate, 100 mL distilled water. To prepare the complete medium, combine 1000 mL broth base, 100 mL $MgCl_2$ solution, and 10 mL malachite green oxalate solution (the total volume of complete medium is 1110 mL). To prepare solution,

dissolve entire contents of $MgCl_2 \cdot 6H_2O$ from newly opened container according to formula, because this salt is very hygroscopic. Malachite green oxalate solution may be stored in dark bottle at room temperature up to 6 months. Merck analytically pure malachite green oxalate is recommended because other brands may not be equally effective. Dispense 10 ml volumes of complete medium into 16 × 150 mm test tubes. Autoclave 15 min at 115 °C. Final pH, 5.5 ± 0.2 (*see* **Note 4**).

9. Xylose lysine desoxycholate (XLD) agar: 3 g yeast extract, 5 g L-lysine, 3.75 g xylose, 7.5 g lactose, 7.5 g sucrose, 2.5 g sodium desoxycholate, 0.8 g ferric ammonium citrate, 6.8 g $Na_2S_2O_3$, 5 g NaCl, 15 g agar, 0.08 g phenol red, 1 L distilled water. Heat with agitation until medium boils. Do not overheat. Pour into plates when medium has cooled down to 50 °C. Let dry about 2 h with covers partially removed. Then close plates. Final pH, 7.4 ± 0.2 (*see* **Note 5**).
10. Hektoen enteric (HE) agar: 12 g peptone, 3 g yeast extract, 9 g bile salts No. 3, 12 g lactose, 12 g sucrose, 2 g salicin, 5 g NaCl, 5 g $Na_2S_2O_3$, 1.5 g ferric ammonium citrate, 0.065 g bromthymol blue, 0.1 g acid fuchsin, 14 g agar, 1 L distilled water. Heat to boiling with frequent agitation to dissolve. Boil no longer than 1 min. Do not overheat. Cool in water bath to 50 °C. Pour 20 mL portions into sterile 15 × 100 mm petri dishes. Let dry 2 h with lids partially open. Final pH, 7.5 ± 0.2 (*see* **Note 6**).
11. Bismuth sulfite (BS) agar: 10 g polypeptone (or peptone), 5 g beef extract, 5 g dextrose, 4 g Na_2HPO_4 (anhydrous), 0.3 g $FeSO_4$ (anhydrous), 8 g bismuth sulfite (indicator), 0.025 g brilliant green, 20 g agar, 1 L distilled water. Mix thoroughly and heat with agitation. Boil about 1 min to obtain uniform suspension. (Precipitate will not dissolve.) Cool down to 45–50 °C. Suspend precipitate by gentle agitation, and pour 20 mL portions into sterile 15 × 100 mm petri dishes. Let plates dry about 2 h with lids partially open, and then close plates. Final pH, 7.7 ± 0.2 (*see* **Note 7**).
12. Triple sugar iron agar (TSI) (Medium 1): 20 g polypeptone, 5 g NaCl, 10 g lactose, 10 g sucrose, 1 g glucose, 0.2 g Fe $(NH_4)_2(SO_4)_2 \cdot 6H_2O$, 0.2 g $Na_2S_2O_3$, 0.025 g phenol red, 13 g agar, 1 L distilled water. Suspend ingredients of Medium 1 in distilled water, mix thoroughly, and heat with occasional agitation. Boil about 1 min to dissolve ingredients. Fill 16 × 150 mm tubes 1/3 full and cap or plug to maintain aerobic conditions. Autoclave Medium 1 for 15 min at 118 °-C. Before the media solidify, incline tubes to obtain 4–5 cm slant and 2–3 cm butt. Final pH, 7.3 ± 0.2 (*see* **Note 8**).

13. Lysine iron agar (LIA) (Edwards and Fife): 5 g gelysate or peptone, 3 g yeast extract, 1 g glucose, 10 g L-lysine hydrochloride, 0.5 g ferric ammonium citrate, 0.04 g $Na_2S_2O_3$ (anhydrous), 0.02 g bromocresol purple, 15 g agar, 1 L distilled water. Heat to dissolve ingredients. Dispense 4 ml portions into 13 × 100 mm screw-cap tubes. Autoclave 12 min at 121 °C. Let solidify in slanted position to form 4 cm butts and 2.5 cm slants. Final pH, 6.7 ± 0.2 (*see* **Note 9**).

3 Methods

In this protocol (Fig. 2), for culture procedures, FDA BAM method (Chapter 5: *Salmonella*) [10] was followed.

3.1 Sample Preparation and Pre-enrichment

1. Twenty-five grams of egg products were aseptically weighed into a Whirl-Pak bag or other appropriate container.
2. Add 225 mL of lactose broth, and mix well by swirling/hand massage (2 min). Let stand for 60 ± 5 min at room temperature (*see* **Note 10**).
3. Mix well by swirling and determine pH with a pH test paper or a pH meter. Adjust pH, if necessary, to 6.8 ± 0.2 with sterile 1 N NaOH and 1 N HCl. Incubate all prepared samples for 24 ± 2 h at 35 °C.

3.2 DNA Extraction

1. Transfer 1 mL pre-enrichment into a 1 mL microcentrifuge tube (*see* **Note 11**). Vortex well.

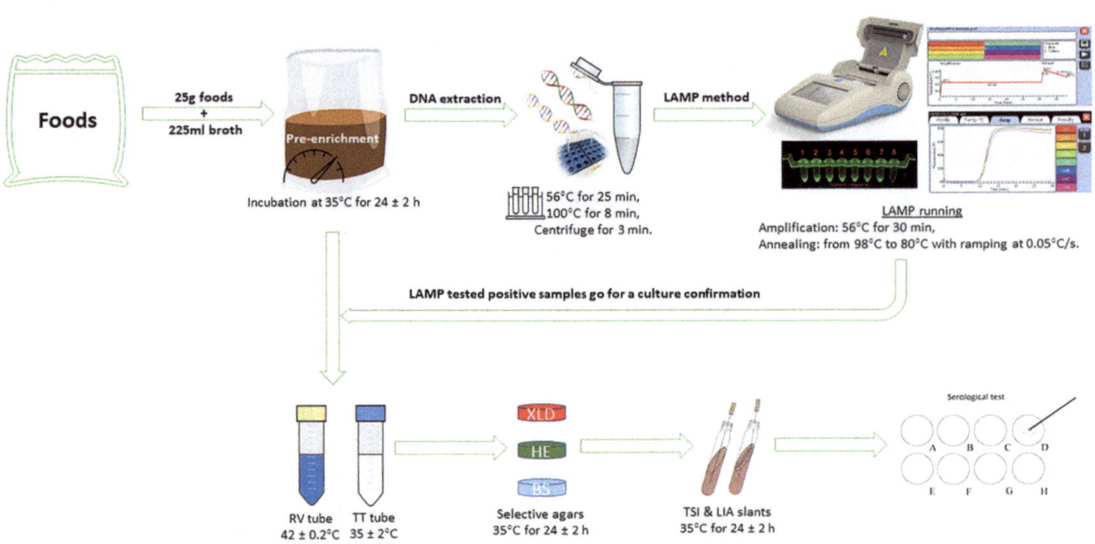

Fig. 2 Work flowchart for detecting *Salmonella* ser. Enteritidis by LAMP method, using the preparation of 25g egg products as an example.

2. Centrifuge the microcentrifuge tube at 12,000 rpm for 1–2 min, and discard the supernatant.
3. Add 200 μL of InstaGene matrix into the pellet and mix well, and incubate at 56 °C for 25 min.
4. Vortex at high speed for 10 s, and heat the tubes at 100 °C for 8 min.
5. Vortex at high speed for 10 s, and spin at 12000 rpm for 3 min.
6. Aliquot the supernatant into a new tube and store at −20 °C or −80 °C for future use.

3.3 LAMP Assay

3.3.1 Instrument and Program Setup

1. Before starting the analysis, please confirm the battery of Genie® III is fully charged or plug the instrument into main electricity outlet and switch on. When running for the first time, check to make sure that the time and date on the status bar are correct. These can be changed from the "Utilities" screen in the toolbox (Fig. 3).
2. Pressing one of the buttons on the main menu will pop up a preview of the profile and allows the user to start, edit, or cancel the run (Fig. 3). The default setting is also the program used for running LAMP. Confirm the program to conduct under isothermal amplification at 65 °C for 30 min followed by annealing (heating and cooling step) from 98 °C to 80 °C with ramping at 0.05 °C/s (*see* **Note 12**). Peak detection threshold of amplification rate and anneal derivative are performed with default setting of 0.010 and 5000, respectively.

3.3.2 LAMP Mix Preparation and Running

1. Thaw the Isothermal Master Mix, primers (*see* Table 1), and prepared DNA extractions at room temperature.
2. Prepare each LAMP reaction mix to a final volume of 25 μL, containing 2.0 μL each of FIP and BIP, 0.5 μL each of F3 and

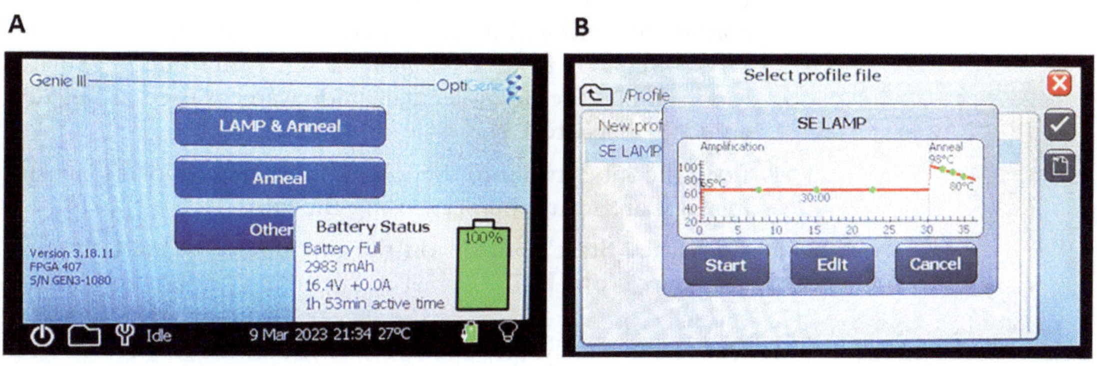

Fig. 3 Battery status and program setting of LAMP running on Genie III. (**a**) battery status; (**b**) program setting

Table 1
Sequences of primers used for *Salmonella* ser. Enteritidis detection by LAMP

Primer	Sequence (5′------- >3′)	Length (bp)
F3	GTGAGGGCGAGGTTTGAAC	19
B3	GGTCATGGTCGGGTATTCGA	20
FIP	TCCTCCACTGACAGGATTCCCA-AAGTGGAGCAGCTGAGCA	40
BIP	CAAACTGAACGTTTGGCTGCCT-GCAGTGACTGGCATCAAGAA	42

F3, forward outer primer; *B3,* backward outer primer; *FIP,* forward inner primer; *BIP,* backward inner primer (*see* **Note 19**).

Table 2
Preparation of LAMP reaction mix

Component	Volume (μL)
Isothermal Mastermix	15
FIP primer (20 pmol)	2.0
BIP primer (20 pmol)	2.0
F3 primer (5 pmol)	0.5
B3 primer (5 pmol)	0.5
DNA template	5.0
Total	**25.0**

F3, forward outer primer; *B3,* backward outer primer; *FIP,* forward inner primer; *BIP,* backward inner primer (*see* **Note 20**).

B3 primers, 5.0 μL of template DNA, and 15.0 μL Isothermal Master Mix (*see* Table 2). Vortex well.

3. For positive controls, include a template DNA prepared from a confirmed *Salmonella* ser. Enteritidis isolate and a template DNA prepared from a *Salmonella* ser. Enteritidis inoculated food sample. Sterile molecular grade water is used as negative control (*see* **Note 13**).
4. Load the tube strip into the LAMP instrument block(s), ensuring caps are secure before closing the lid.
5. Click the "Start" button on the instrument to start the program (*see* **Note 14**).

3.3.3 Data Analysis and Results Calling

1. Data can be analyzed directly at the end of the LAMP run or at a later time by opening the stored data file from both the Genie® III device and a laptop connect to the device installed with a software named Genie Explorer software [11].

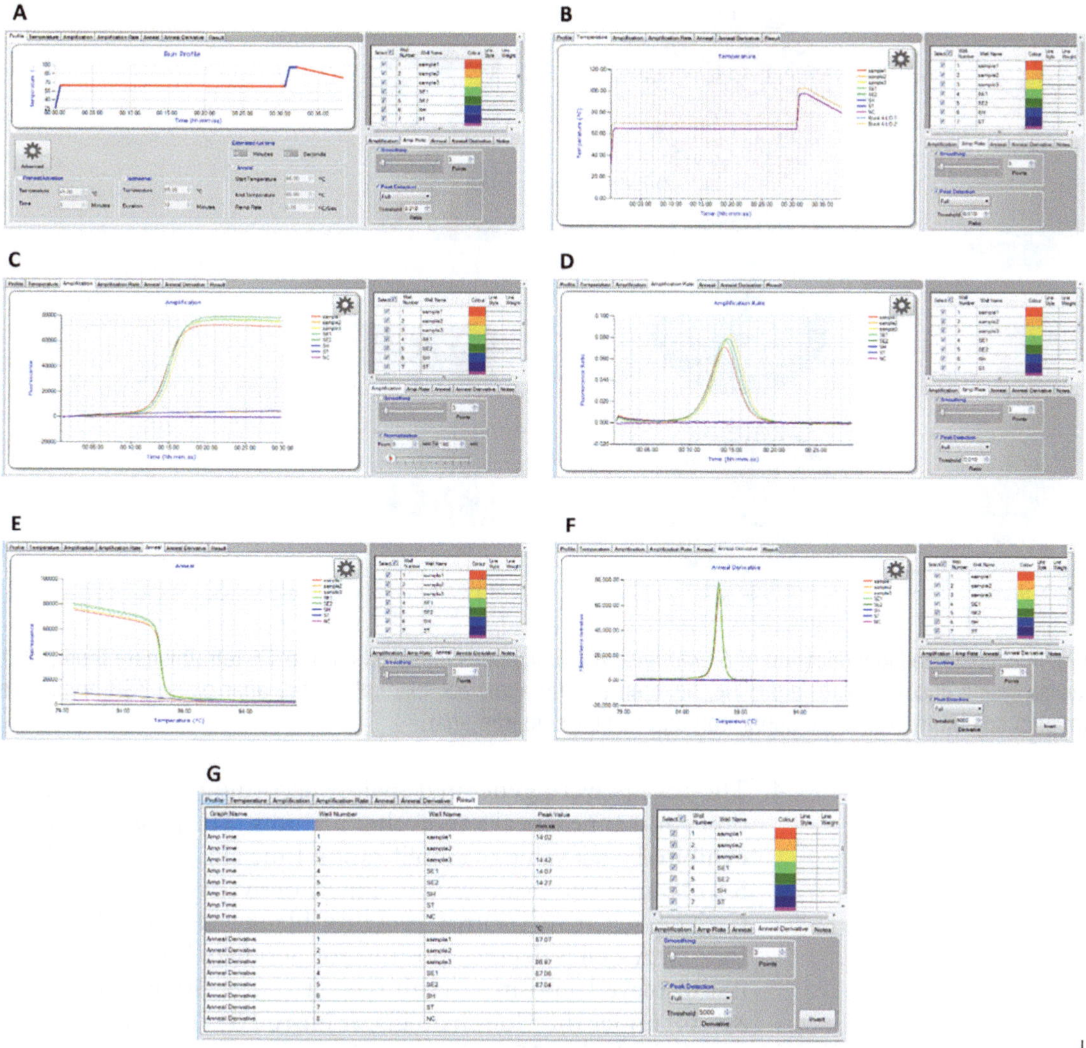

Fig. 4 Results of LAMP running on Genie III. (**a**) run profile setting; (**b**) temperature interpretation; (**c**) amplification results; (**d**) amplification rate; (**e**) annealing results; (**f**) annealing derivative; (**g**) results display

2. Once the data analysis parameters have been set, results are interpreted for each sample by analyzing the fluorescence data that is being acquired during the amplification phase of the experiment and the fluorescence derivative data that is being acquired during the anneal phase of the experiment. Click the Well Name to check the results for each well (Fig. 4).

3. Click the "Results" button on the screen to check the details of amplification time and annealing temperature. Pressing the button shown on the results page will also help to generate a PDF report with the amplification graph, the anneal graph, and the results table. The button is for generating a CVS file and for a direct printing.

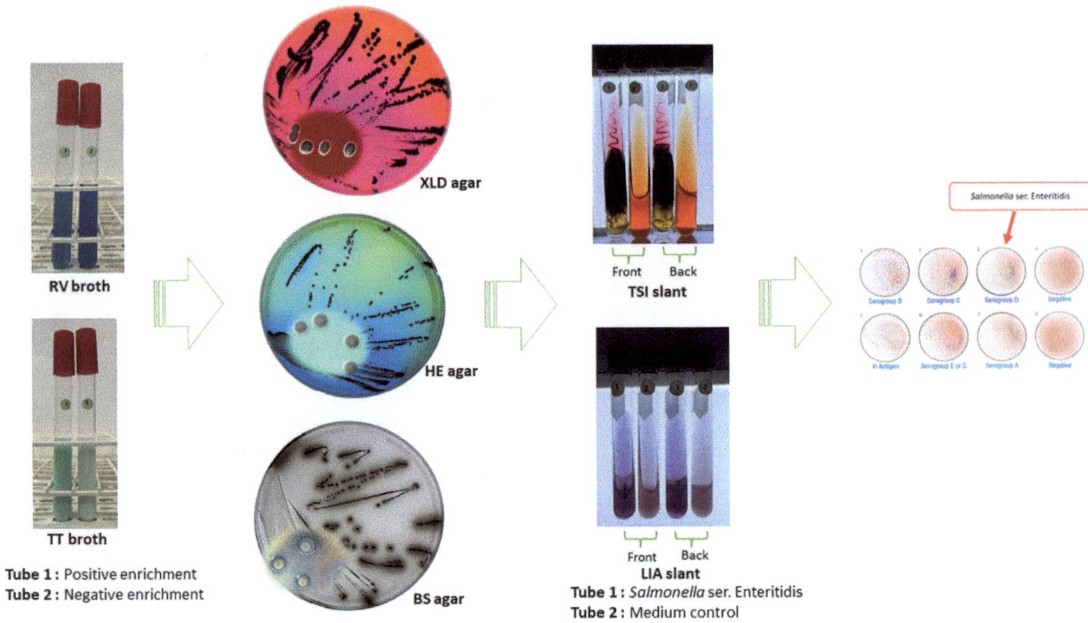

Fig. 5 Isolation and identification of *Salmonella. RV,* Rappaport-Vassiliadis broth; *TT,* tetrathionate broth; *XLD,* xylose lysine desoxycholate agar; *HE,* hektoen enteric agar; *BS,* bismuth sulfite agar plate; *TSB,* trypticase (tryptic) soy broth; *TSI,* triple sugar iron agar; *LIA,* lysine iron agar.

4. The results were generated mainly according to amplification times and anneal peaks and other features. The positive control exhibited peak value of amplification time (P_{Amp}) ranging from 5 to 15 min and peak value of anneal derivative (P_{Anneal}) ~ 87 °C. The negative control presents blank value of P_{Amp} and P_{Anneal}. A well is deemed positive for *Salmonella* ser. Enteritidis if it displayed a positive result type (P_{Amp}: 5–30 min, P_{Anneal}: ~ 87 °C) under the blue channel (*see* **Note 15**). The run is usually terminated after 35 min of reaction.
5. If a sample is found positive for *Salmonella*, proceed to the following procedures for culture confirmation and identification of the suspected isolates (Fig. 5). If negative, a decision to proceed to the following steps or not will depend on the analyst's needs.

3.4 Salmonella *Isolation and Identification*

1. Tighten bag/lid and gently shake incubated sample. Transfer 0.1 and 1.0 mL of incubated sample to 10 mL RV broth and 10 mL TT broth, respectively. Vortex well and then incubate RV tubes at 42 ± 0.2 °C for 24 ± 2 h and TT tubes at 35 ± 2 °C for 24 ± 2 h.
2. Vortex RV and TT tubes thoroughly, and streak 1 loopful (10 μL) onto each of the XLD, HE, and BS agar plates (usually

with duplicates of each type). Incubate all plates at 35 ± 2 °C for 24 ± 2 h.

3. Examine plates for presence of colonies that may be *Salmonella*.
4. For typical *Salmonella* colony (*see* **Note 16**), pick two or more colonies from each selective agar plate, and then inoculate onto TSI and LIA slants. Lightly touch the very center of the colony to be picked with sterile inoculating needle and inoculate TSI slant by streaking slant and then stabbing butt. Without flaming, inoculate LIA slant by stabbing butt twice and then streaking slant (*see* **Note 17**).
5. If typical colonies are present on the BS agar after 24 ± 2 h incubation, then pick two or more colonies. Irrespective of whether or not BS agar plates are picked at 24 ± 2 h, reincubate BS agar plates for additional 24 ± 2 h. After a total of 48 ± 2 h incubation, pick two or more typical colonies, if present, from the BS agar plates, only if colonies picked from the BS agar plates incubated for 24 ± 2 h give atypical reactions in TSI and LIA that result in culture being discarded as not being *Salmonella*.
6. In the absence of typical *Salmonella* colonies on HE or XLD agars after 24 ± 2 h incubation, pick two or more atypical *Salmonella* colonies and inoculate onto TSI and LIA slants. If typical or suspicious colonies are not present on BS agar after 24 ± 2 h, then do not pick any colonies. Reincubate BS plates for additional 24 ± 2 h. If typical or suspicious colonies are not present after 48 ± 2 h incubation, then pick two or more atypical colonies, and inoculate onto TSI and LIA slants.
7. Incubate all TSI and LIA slants at 35 ± 2 °C for 24 ± 2 h (*see* **Note 18**).
8. Selected typical *Salmonella* colonies (2 colonies/plate) should be confirmed with Wellcolex™ Color *Salmonella* Rapid Latex Agglutination Test Kit (Remel Europe Ltd., CA, USA) or other validated confirmation methods.

4 Notes

1. For more details, see BAM Media M154: Trypticase (Tryptic) Soy Broth, https://www.fda.gov/food/laboratory-methods-food/bam-media-m154-trypticase-tryptic-soy-broth
2. For more details, see BAM Media M74: Lactose Broth, https://www.fda.gov/food/laboratory-methods-food/bam-media-m74-lactose-broth
3. Do not autoclave tetrathionate (TT) broth base, and heat the mixed broth base to boiling. The precipitate will not dissolve

completely. Do not heat medium after addition of I_2-KI and dye solutions. See BAM Media M145: Tetrathionate (TT) Broth, https://www.fda.gov/food/laboratory-methods-food/bam-media-m145-tetrathionate-tt-broth

4. When preparing Rappaport-Vassiliadis (RV) broth, the broth base must be prepared on the same day that components are combined to make complete medium. Magnesium chloride solution may be stored in dark bottle at room temperature up to 1 year. Store the prepared RV broth in refrigerator and use within 1 month. For more details, see BAM Media M132: Rappaport-Vassiliadis Medium, https://www.fda.gov/food/laboratory-methods-food/bam-media-m132-rappaport-vassiliadis-medium
5. Heat with agitation until medium boils. Do not overheat. XLD agar may be stored up to 30 days under refrigeration (4 ± 2 °C). For more details, see BAM Media M179: Xylose Lysine Desoxycholate (XLD) Agar, https://www.fda.gov/food/laboratory-methods-food/bam-media-m179-xylose-lysine-desoxycholate-xld-agar
6. Heat to boiling with frequent agitation to dissolve. Boil no longer than 1 min. Do not overheat. HE agar may be stored up to 30 days under refrigeration (4 ± 2 °C). For more details, see BAM Media M61: Hektoen Enteric (HE) Agar, https://www.fda.gov/food/laboratory-methods-food/bam-media-m61-hektoen-enteric-he-agar
7. Do not autoclave BS agar. Prepare BS plates the day before use and store in dark at room temperature until streaked. Selectivity decreases in 48 h. For more details, see BAM Media M19: Bismuth Sulfite Agar (Wilson and Blair), https://www.fda.gov/food/laboratory-methods-food/bam-media-m19-bismuth-sulfite-agar-wilson-and-blair
8. For the preparation of triple sugar iron agar (TSI), two options (Medium 1 and Medium 2) are interchangeable for general use; we selected Medium 1 in this protocol. For more details, see BAM Media M149: Triple Sugar Iron Agar (TSI), https://www.fda.gov/food/laboratory-methods-food/bam-media-m149-triple-sugar-iron-agar-tsi
9. For more details, see BAM Media M89: Lysine Iron Agar (Edwards and Fife), https://www.fda.gov/food/laboratory-methods-food/bam-media-m89-lysine-iron-agar-edwards-and-fife
10. Select matched broth for different food matrices following FDA BAM Chapter 5: *Salmonella*. For example, lactose broth is used for dried egg yolk, dried egg whites, and dried whole eggs; TSB is used for shell eggs and hard-boiled eggs; and TSB

supplemented with ferrous sulfate is used for liquid whole eggs (homogenized).

11. The positive control for *Salmonella* ser. Enteritidis can be prepared by either adding 2–3 colonies into 200 μL of InstaGene matrix or by centrifuging 1 mL of pure culture and then adding 200 μL of InstaGene matrix. Both methods are viable for preparation. Inoculate the sample intended for testing as the inoculated positive control. When dealing with extremely low-contamination samples, it is advisable to prepare a 2 ml pre-enrichment solution for DNA extraction. This larger volume ensures an adequate amount of material for extraction, resulting in both higher yields and better DNA quality.
12. On Genie® III, there are two ways to start a run: by touching one of the predefined profiles saved on the instrument (or an attached USB pen drive) or by creating a new profile. Predefined profiles that are saved on the instruments internal memory can be selected for quick start.
13. It is highly recommended to load negative control first and positive control at the end to avoid misplacement and cross-contamination.
14. It is strongly recommended to avoid opening the reaction tubes after the program is completed. However, if it becomes necessary to open them for running gels, it is crucial to physically separate the areas used for running LAMP and gels, as well as for preparing DNA extraction. This separation helps prevent aerosol contamination that may occur due to the opening of the tubes.
15. It is highly recommended to conduct duplicate runs. The final result should be generated by accumulating consistent data obtained from these repeated runs, ensuring reliability and accuracy in the analysis. If the results of two or more replicates do not match or curve of the run appears very late (e.g., >25 min), consider re-run the tests. If you suspect the template DNA quality is insufficient, you can re-extract DNA and run the assay again.
16. Typical *Salmonella* colonies on HE agar are blue-green to blue with or without black centers. Many cultures of *Salmonella* may produce colonies with large, glossy black centers or may appear as almost completely black. Typical *Salmonella* colonies on XLD agar are pink with or without black centers. Many cultures of *Salmonella* may produce colonies with large, glossy black centers or may appear as almost completely black. Typical *Salmonella* colonies on BS agar are brown, gray, or black; sometimes they have a metallic sheen. Surrounding medium is usually brown at first, but may turn black in time with increased incubation, producing the so-called halo effect.

17. Since lysine decarboxylation reaction is strictly anaerobic, the LIA slants must have deep butt (4 cm).
18. Cap tubes loosely to maintain aerobic conditions while incubating slants to prevent excessive H_2S production. *Salmonella* typically produces alkaline (red) slant and acid (yellow) butt, with or without production of H_2S (blackening of agar) in TSI. In LIA, *Salmonella* typically produces alkaline (purple) reaction in butt of tube. Consider only distinct yellow in butt of tube as acidic (negative) reaction. Do not eliminate cultures that produce discoloration in butt of tube solely on this basis. Most *Salmonella* cultures produce H_2S in LIA. Some non- *Salmonella* cultures produce a brick-red reaction in LIA slants.
19. The primer sets were designed with PrimerExplorer V4 software and synthesized by Integrated DNA Technologies, Inc. (Coralville, Iowa, USA). Rehydrate each primer with sterile molecular grade water to a working solution of 10 μM following the manufacturer's instruction. Vortex well and store at −20 °C / −80 °C for further use.
20. It is highly recommended to calculate and prepare a working solution based on the targeted number of samples. Additionally, it is important to utilize a clean area for preparing the reaction mix before adding the DNA template. This approach ensures accuracy and minimizes the risk of contamination during the experimental process.

Acknowledgments

This work was supported by funding from the Office of Laboratory Operations and Applied Science, Human Foods Program, U.S. Food and Drug Administration.

Disclaimer Mention of trade names or commercial products in the paper is solely for the purpose of providing scientific information and does not imply recommendation or endorsement by the US Food and Drug Administration. This is a scientific publication. This publication represents the views of the authors and does not represent the views of the U.S. Food and Drug Administration.

References

1. Notomi T, Okayama H, Masubuchi H, Yonekawa T, Watanabe K, Amino N et al (2000) Loop-mediated isothermal amplification of DNA. Nucleic Acids Res 28:e63
2. Eiken GENOME SITE (2021) Basic principle of LAMP method (Animation). https://loopamp.eiken.co.jp/en/lamp/0211.html. Accessed 8 Jan 2024
3. Tomita N, Mori Y, Kanda H, Notomi T (2008) Loop-mediated isothermal amplification (LAMP) of gene sequences and simple visual detection of products. Nat Protoc 3:877–882

4. Eiken Chemical Co. L (2023) LAMP basic principle. https://loopamp.eiken.co.jp/en/lamp/0203.html. Accessed 9 Jan 2024
5. Francois P, Tangomo M, Hibbs J, Bonetti E-J, Boehme CC, Notomi T et al (2011) Robustness of a Loop-mediated isothermal amplification reaction for diagnostic applications. FEMS Immunol Med Microbiol 62:41–48
6. Kaneko H, Kawana T, Fukushima E, Suzutani T (2007) Tolerance of Loop-mediated isothermal amplification to a culture medium and biological substances. J Biochem Biophys Methods 70:499–501
7. Becherer L, Borst N, Bakheit M, Frischmann S, Zengerle R, von Stetten F (2020) Loop-mediated isothermal amplification (LAMP)–review and classification of methods for sequence-specific detection. Anal Methods 12:717–746
8. Martelli F, Davies RH (2012) *Salmonella* serovars isolated from table eggs: an overview. Food Res Int 45:745–754
9. OptiGene (2023) GspSSD LF DNA polymerase. http://www.optigene.co.uk/gspssd-lf-dna-polymerase/. Accessed 11 Oct 2023
10. U.S. food and Drug Administration (FDA) (2023) Chapter 5: *Salmonella*. https://www.fda.gov/food/laboratory-methods-food/bam-chapter-5-salmonella. Accessed 8 Jan 2024
11. OptiGene (2021) Genie III User manual. http://www.optigene.co.uk/wp-content/uploads/2021/01/Genie%20III%20Manual%20version%201.11.pdf. Accessed 12 Jan 2024
12. Huang T, Li L, Liu X, Chen Q, Fang X, Kong J et al (2020) Loop-mediated isothermal amplification technique: principle, development and wide application in food safety. Anal Methods 12:5551–5561

Part III

Virus

Chapter 10

One-Step Reverse Transcription Loop-Mediated Isothermal Amplification of Norovirus Genogroup I and II in Oysters

Soontag Jung, Daseul Yeo, Hyojin Kwon, and Changsun Choi

Abstract

A one-step reverse transcription loop-mediated isothermal amplification (RT-LAMP) was developed for the rapid and sensitive detection of human norovirus (HuNoV) in oysters. HuNoV, a leading cause of foodborne gastroenteritis, is frequently associated with contaminated oysters, necessitating a reliable on-site detection method. RT-LAMP was designed using primers targeting conserved regions of HuNoV genogroups I (GI) and II (GII). This method demonstrated a detection limit of 10^1 genome copies per microliter, surpassing the sensitivity of semi-nested RT-PCR and real-time RT-PCR. The restriction fragment length polymorphism analysis confirmed specificity, ensuring no cross-reactivity with other foodborne pathogens. RT-LAMP reaction was optimized for isothermal amplification at 63 °C, completed within 60 min. Viral RNA was extracted using the QIAamp Viral RNA Mini Kit, and results were visualized through agarose gel electrophoresis or colorimetric detection with SYBR Green I dye. Field tests using artificially contaminated oyster samples confirmed the reliability and accuracy of RT-LAMP, making it suitable for on-site applications in the oyster industry. This study highlights the potential of one-step RT-LAMP as a practical, rapid, and sensitive diagnostic tool for HuNoV detection. Its simplicity, cost-effectiveness, and ability to deliver results without sophisticated equipment make it an ideal solution for enhancing food safety and preventing HuNoV outbreaks. Such advancements are anticipated to strengthen the global competitiveness of the oyster industry by ensuring the microbiological safety of seafood products.

Key words Human norovirus (HuNoV), Loop-mediated isothermal amplification (LAMP), Food-borne virus, Oyster, Hemocytes, On-site detection

1 Introduction

Human norovirus (HuNoV) is recognized as one of the leading causes of acute viral gastroenteritis worldwide, significantly impacting public health and food safety [1]. Belonging to the *Caliciviridae* family, noroviruses are classified into ten genogroups (GI–GX), with genogroups I (GI) and II (GII) being most commonly associated with human infections [2]. The virus is highly contagious and is primarily transmitted via the fecal-oral route, through

Alejandro Garrido-Maestu and Alexandre Lamas (eds.), *Foodborne Pathogens: Loop-mediated Isothermal Amplification Protocols*, Methods and Protocols in Food Science, https://doi.org/10.1007/978-1-0716-4526-0_10,

contaminated food or water, and by person-to-person contact. Additionally, airborne transmission has been reported in crowded settings [3]. HuNoV infections are characterized by symptoms such as abdominal pain, diarrhea, nausea, vomiting, and mild fever, with an incubation period ranging from 10 to 51 h [4]. While the illness is generally self-limiting in healthy individuals, it poses severe risks to vulnerable populations, including young children, the elderly, and immunocompromised individuals [2].

Among foodborne transmission routes, the consumption of raw or undercooked oysters has been identified as a major contributor to HuNoV outbreaks. As filter feeders, oysters accumulate viral particles from contaminated seawater in their digestive tissues, making them a reservoir for HuNoV [5]. This presents a significant challenge for the oyster industry, as HuNoV contamination threatens public health and leads to substantial economic losses due to trade restrictions and recalls. For instance, outbreaks linked to HuNoV-contaminated oysters have led to temporary bans on exports from major oyster-producing countries, including South Korea, to markets such as the United States and New Zealand [6].

Given the public health and economic implications, the development of rapid, sensitive, and reliable methods for detecting HuNoV in oysters is critical. Traditional detection methods, such as enzyme-linked immunosorbent assays and polymerase chain reaction (PCR)-based techniques, are widely used but often require sophisticated equipment, trained personnel, and prolonged processing times, limiting their application in field settings [7].

Therefore, this study developed one-step reverse transcription loop-mediated isothermal amplification (RT-LAMP) for detecting HuNoV GI and GII in oysters. RT-LAMP offers several advantages, including high sensitivity, specificity, and rapid amplification under isothermal conditions, making it suitable for on-site testing. This chapter explores the principles of RT-LAMP, its application in detecting HuNoV in oysters, and its potential to enhance food safety and support the competitiveness of the oyster industry in global markets.

2 Materials

2.1 Standard HuNoV RNAs and Stool Samples

1. Synthetic standard RNA for HuNoV GI (VR-3234SD) and GII (VR-3235SD) were purchased from the American Type Culture Collection (ATCC; Manassas, VA, USA).
2. Stool samples of diarrheal patients infected with HuNoV were provided by the Korea Center for Disease Control and Prevention (KCDC; Cheongwon, Chungcheongbuk, Korea). The samples were diluted to a 1:10 ratio using sterile phosphate-buffered saline (PBS, pH 7.4).

2.2 Nucleic Acids for Specificity Testing

1. Nucleic acids from the following viruses were used for specificity testing:
   - Hepatitis A virus (HAV, ATCC VR-1402).
   - Hepatitis E virus (HEV, stool isolate).
   - Rotavirus A Wa strain (ATCC VR-2018).
   - Human adenovirus 41 (ATCC VR-930).
2. Positive controls included HuNoV GI and GII RNA confirmed by nested RT-PCR and sequence analysis.

2.3 RNA Isolation

1. QIAamp Viral RNA Mini Kit from Qiagen involved in obtaining viral RNA from oyster samples and control as an extraction.

2.4 Loop-Mediated Isothermal Amplification (LAMP) Reagents

1. 10 × ThermoPol buffer (New England Biolabs, Ipswich, MA, USA)
2. 5 × cDNA synthesis buffer (Invitrogen, Carlsbad, CA, USA)
3. Cloned avian myeloblastosis virus reverse transcriptase (15 U, Invitrogen).
4. *Bst* DNA polymerase (8 U, New England Biolabs).
5. Loopamp RNA amplification kit (Eiken Chemical, Tochigi, Japan).
6. 10 mM dNTPs (Bioneer, Daejeon, Korea)
7. SYBR Green I dye (Invitrogen) for visual detection of LAMP products.

2.5 PCR Reagents

1. 10 × PCR buffer (Bioneer)
2. Oligo d(T) primer (50 mM) (Invitrogen).
3. RNase inhibitor (20 U/mL) (Invitrogen).
4. Murine leukemia virus reverse transcriptase (50 U/mL) (Invitrogen).
5. Premix Ex Taq (2×) buffer (TaKaRa Bio Inc., Shiga, Japan) for real-time RT-PCR.
6. Semi-nested RT-PCR and real-time RT-PCR primer sets (Table 1).

2.6 Restriction Enzymes for Specificity Testing

1. *Hha*I and *Hae*III restriction enzymes for restriction fragment length polymorphism (RFLP) analysis.
2. NEB Cutter V2.0 software was used to select restriction enzymes.

Table 1
Primers and probe sets of snRT-PCR, real-time RT-PCR, and RT-LAMP used for the detection of HuNoV GI and GII

HuNoV detection		Primer/probe	Region	Oligonucleotide sequence (5′-3′)	References
HuNoV GI	snRT-PCR	GI-F1	ORF1–ORF2	5′-CTGCCCGAATTYGTAAATGATGAT-3′	[8]
		GI-R1	ORF2	5′-CCAACCCARCCATTRTACATYTG-3′	
		GI-F2	ORF1–ORF2	5′-ATGATGATGGCGTCTAAGGACGC-3′	
	Real-time RT-PCR	COG1F	ORF1	5′-CGYTGGATGCGNTTYCATGA-3′	[9]
		COG1R	ORF1–ORF2	5′-CTTAGACGCCATCATCATTYAC-3′	
		RING1(a)-TP	ORF1	5′-AGATYGCGATCYCCTGTCCA-3′	
		RING1(b)-TP	ORF1	5′-AGATCGCGGTCTCCTGTCCA-3′	
	Real-time RT-LAMP	F3	ORF1	5′-CCRGGNTGGCARGCNATGTT-3′	[10]
		B3	ORF2	5′-CCAACCCARCCATTRTACA-3′	
		FIP1	ORF1–ORF2	5′-CATTTACGAATTCGGGCAGGTTTTCGCTGGATGCGNTT CCATGA-3′	
		FIP2	ORF1–ORF2	5′-CATTTACAAAATCGGGCAGGTTTTCGCTGGATGCGNTT CCATGA-3′	
		BIP1	ORF1–ORF2	5′-GATGGCGTCTAAGGACGCTTTTAGCTGTRTTTGCCTCTG GWAC-3′	
		BIP2	ORF1–ORF2	5′-GATGGCGTCTAAGGACGCTTTTAGCWGTATTAACCTCC GGYAC-3′	
		LF1	ORF1	5′-AGATYGCGATCYCCTGTCCA-3′	
		LF2	ORF1	5′-AGATTGCGATCTCCTGCCCA-3′	
		LF3	ORF1	5′-AGCTCGCGGTCTCCTGTCCA-3′	
	One-step RT-LAMP	F3 779M	ORF1	5′-YATGTTCCGYTGGATGCG-3′	[11]
		B3 982	ORF2	5′-AACTTGCCCAGCAGTTGC-3′	

		FIP 805	ORF1–ORF2	5′-AGCGTCCTTAGACGCCATCATCACCTCGGATTGTGGACAGG-3′	
		BIP 901	ORF2	5′-GGCGCTGGTCAGTTGGTACCCGCTACAGGATCCATTGCA-3′	
HuNoV GII	snRT-PCR	GII-F1	ORF1	5′-GGGAGGGCGATCGCAATCT-3′	[8]
		GII-R1	ORF2	5′-CCRCCIGCATRICCRTTRTACAT-3′	
		GII-F2	ORF1	5′-TTGTGAATGAAGATGGCGTCGART-3′	
	Real-time RT-PCR	COG2F	ORF1	5′-CARGARBCNATGTTYAGRTGGATGAG-3′	[9]
		COG2R	ORF1–ORF2	5′-TCGACGCCATCTTCATTCACA-3′	
		RING2-TP	ORF1	5′-TGGGAGGGCGATCGCAATCT-3′	
	Real-time RT-LAMP	F3	ORF1	5′-GGNMTGGANTTTTAYGTGCCMAG-3′	[10]
		B3	ORF2	5′-CCRCCNGCATRHCCRTTRTACAT-3′	
		FIP1	ORF1	5′-GGGAGCMAGATTGCGATCGCTTTTGAGBCNATGTTYAGRTGGAT-3′	
		FIP2	ORF1	5′-GGGAGCMAGATTGCGATCGCTTTTGAGCCCATGTTCAGRTGGAT-3′	
		LF3	ORF1	5′-GTGCTCAAATCTGAGAATCTC-3′	
		LF4	ORF1	5′-GTGCTÇAAGTCTGAGAAYCTC-3′	
	One-step RT-LAMP	F3 33	ORF2	5′-CCCATCTGATGGGTCCRCA-3′	[11]
		B3 206	ORF2	5′-CACCTGGAGCGTTTCTAGG-3′	
		FIP 97	ORF2	5′-ATAGCGGCACCAACAACGGCCTCGTCCCAGAGGTCAAC-3′	
		BIP 122	ORF2	5′-ACCTGTAGCGGGCCAACAACTCTCCACCAGGGGCTT-3′	
		FIP 45	ORF1	5′-ACCTTTTGGGCCAGATGCGGCGTGTTAGCAGCAGCCTT-3′	
		BIP 121	ORF1–ORF2	5′-CTTCTCTCTCCCAGCGTGATGGGGGGAGACCCTCATTGATTG-3′	

ORF open reading frame, *snRT-PCR* Semi-nested Reverse Transcription-PCR

2.7 Oyster Samples

1. Medium-sized pacific oysters (*Crassostrea gigas*) were purchased from a retail market in Tongyeong, South Korea.
2. Oysters were acclimated in natural seawater at 15 °C for 1 week before viral contamination experiments.
3. Alternatively, 5–20 mL of hemocytes and hemolymph were harvested by shucking and slicing 4–5 oysters. The viral RNA was extracted by following the previous protocol [12].

2.8 Consumables

1. Sterile phosphate-buffered saline (PBS, pH 7.4) for sample preparation.
2. DEPC-treated water for RNA handling.
3. 1.5% agarose gel for electrophoresis
4. Agarose powder (Promega, Madison, WI, USA).
5. Ethidium bromide (EtBr) for gel staining.
6. PCR tube for real-time RT-PCR.

2.9 Equipment

1. CFX96 Touch Real-Time PCR Detection System (Bio-Rad, Hercules, CA, USA).
2. T100 thermal cycler (Bio-Rad) for LAMP and semi-nested RT-PCR.
3. Gel electrophoresis apparatus and power supply (Bio-Rad).
4. Epoch (BioTek, Winooski, VT, USA) for RNA quantification.
5. Benchtop centrifuge for RNA extraction and sample preparation.
6. Adjustable micropipettes (Eppendorf).
7. Sterile, DNase/RNase-free pipette tips.
8. Loopamp real-time turbidimeter (LA-500; Teramecs, Kyoto, Japan).

3 Methods

3.1 Sample Collection and Preparation

1. Stool samples from diarrheal patients infected with HuNoV were provided by the KCDC. The samples were diluted 1:10 with sterile phosphate-buffered saline (PBS, pH 7.4) and centrifuged at 10,000× *g* for 5 min to remove debris. The supernatant was used for RNA extraction.
2. Pacific oysters (*Crassostrea gigas*) were purchased from a retail market in Tongyeong, South Korea, and acclimated in HuNoV-free natural seawater at 15 °C for 1 week. For contamination, 1 × 10^6 copies of HuNoV GI.1 were added to

500 mL of seawater in a 1-L beaker at 15 °C. Six oysters were incubated individually in the contaminated seawater overnight at 15 °C.

3. For hemocyte analysis, hemocytes and hemolymph were collected by shucking the oysters and slicing the oyster meat 4–5 times. Approximately 5–20 mL of hemolymph containing hemocytes was drained into a tube.
4. For digestive gland analysis, 3 g of tissue was excised and homogenized with 3 mL of 100 mg/mL proteinase K solution. The homogenate was incubated at 37 °C for 60 min and then heated at 60 °C for 15 min to deactivate the enzyme. After centrifugation at 4000× *g* for 5 min, the supernatant was transferred to a new tube and centrifuged again.

3.2 RNA Extraction

1. Viral RNA was extracted from NoV-positive stools, oyster homogenates, oyster hemocytes, and viral culture supernatants using the QIAamp Viral RNA Mini Kit (Qiagen) according to the manufacturer's protocol.
2. The extracted RNA was eluted in 50 μL of nuclease-free water and stored at −80 °C until use. RNA concentration and purity (A260/A280 ratio) were determined using an Epoch (BioTek).

3.3 Primer Design

1. Primers for RT-LAMP were designed to specifically target the highly conserved regions of the HuNoV genome [11]. The following steps were undertaken:
2. The ORF1–ORF2 junction region of the HuNoV genome was chosen as the target for primer design due to its high conservation across genogroups GI and GII.
3. Reference sequences for HuNoV GI and GII were retrieved from the GenBank database. Multiple sequence alignments were performed using ClustalW to identify conserved regions suitable for primer binding.
4. LAMP primers were designed using the Primer Explorer V4 software (http://primerexplorer.jp/elamp4.0.0/index.html).
5. The primer set included six primers targeting eight distinct regions of the conserved viral RNA sequence (Fig. 1).
6. Primer sequences were evaluated for melting temperature (Tm), GC content, and secondary structure using the Primer Explorer V4 tool. The Tm of the inner primers was optimized to be approximately 60–65 °C, while the Tm of the outer primers was set to around 55–60 °C.

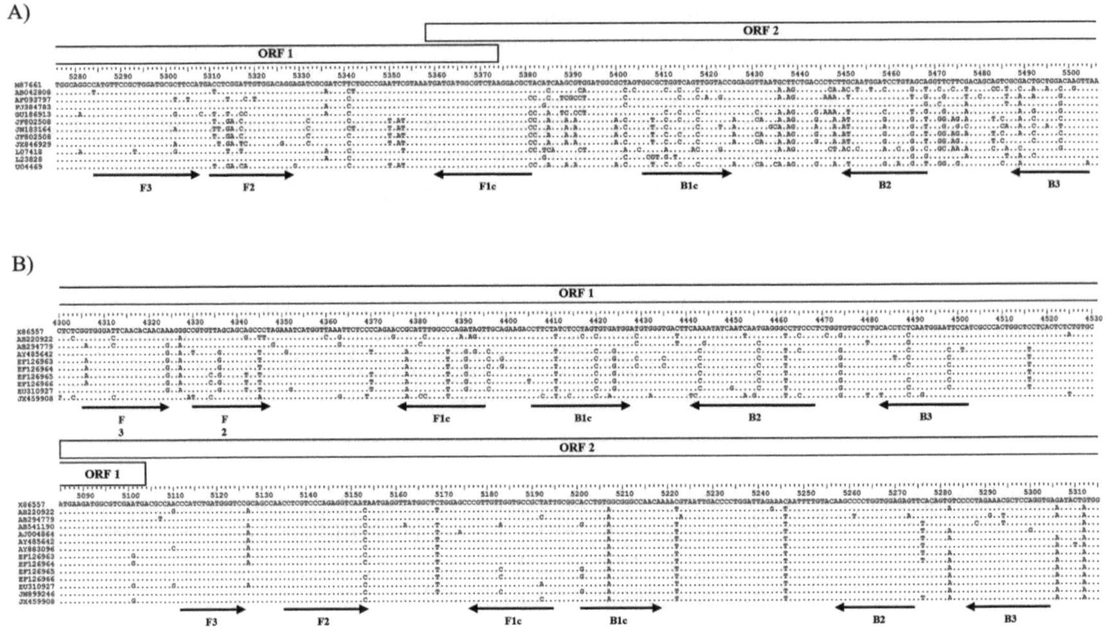

Fig. 1 Design of one-step RT-LAMP primer set on the highly homologous sequences of human norovirus GI and GII. (**a**) One-step RT-LAMP primers for detecting HuNoV GI based on GenBank accession no. FJ384783. (**b**) Primer binding sites of HuNoV GII on nucleotide sequence data for GenBank accession no. X86557. There are two outer primers (F3, B3) and two forward and backward inner (FIP, BIP) primers. The inner primer FIP or BIP are composed of F2 and F1c or B2 and B1c, respectively. (Permitted from Jeon et al. [11])

3.4 One-Step RT-LAMP

1. LAMP primers for HuNoV detection were designed based on the highly conserved ORF1–ORF2 regions of prevalent HuNoV GI and GII (Table 1) [11].
2. For one-step RT-LAMP, the reaction was conducted in a 25 μL mixture containing 2.5 μL of 10 × Thermopol buffer (New England Biolabs), 1.35 μL of 5 × cDNA synthesis buffer (Invitrogen), 2 μL of 10 mM dNTPs (Bioneer), 0.5 μL of 15 U cloned avian myeloblastosis virus reverse transcriptase (Invitrogen), 1 μL of 8 U *Bst* DNA polymerase (New England Biolabs), 1 μL of 10 pM outer primers F3 and B3 (specific sets for HuNoV GI or GII), 4 μL of 20 pM inner primers FIP and BIP (specific sets for HuNoV GI or GII), 2 μL of RNA template, and 5.65 μL of DEPC-treated water.
3. The reaction mixture was incubated at 63 °C for 60 min, followed by inactivation at 80 °C for 2 min.
4. The LAMP products were analyzed on a 1.5% agarose gel using gel electrophoresis. Additionally, SYBR Green I dye (0.5 μL) was added to the reaction products for colorimetric detection,

where positive reactions turned green under the UV illuminator, while negative reactions remained orange.

3.5 Real-Time RT-LAMP

1. A broadly reactive real-time RT-LAMP assay for HuNoV detection was performed using the Loopamp RNA Amplification Kit (Eiken Chemical) [10].
2. The reaction mixture contained 12.5 μL of 2 × Loopamp reaction mix; 1 μL of Loopamp enzyme mix; 5 pM of outer primers F3 and B3; 25 pM of inner primers FIP1, FIP2, BIP1, and BIP2 for HuNoV GI; or 40 pM of FIP1 and BIP1 and 20 pM of FIP2, FIP3, BIP2, BIP3, and BIP4 for HuNoV GII. Additionally, 20 pM of loop primers LF1, LF2, and LF3 were included for HuNoV GI, whereas 20 pM of LF1, LF2, LF3, and LF4 were used for HuNoV GII (Table 1). Finally, 2 μL of RNA extract was added to the reaction mixture.
3. The reaction was incubated at 62 °C for 90 min, followed by heating at 80 °C for 5 min, using a Loopamp real-time turbidimeter (Teramecs).
4. Amplification was monitored in real time, and the specificity of the assay was confirmed through the use of highly specific primers and optimized reaction conditions.

3.6 Semi-nested Reverse Transcription-PCR (snRT-PCR)

1. Reverse transcription (RT) was performed in a total reaction volume of 10 μL, consisting of 1 μL of 10× PCR buffer (Bioneer), 1 μL of 10 mM dNTP (Bioneer), 0.5 μL of 50 mM oligo d(T) (Invitrogen), 0.5 μL of 20 U/μL RNase inhibitor (Invitrogen), 0.5 μL of 50 U/μL murine leukemia virus reverse transcriptase (Invitrogen), 1 μL of target RNA, and 5.5 μL of DEPC-treated water. RT was conducted using a T100 thermal cycler (Bio-Rad) at 42 °C for 15 min, followed by denaturation at 99 °C for 5 min.
2. The first round of PCR was performed in a total reaction volume of 20 μL, which included 2 μL of 10× PCR buffer (Bioneer), 2 μL of 10 mM dNTP (Bioneer), 1 μL of 20 pM GI-F1 and GI-R1 primers for HuNoV GI or 1 μL of 20 pM GII-F1 and GII-R1 primers for HuNoV GII (Table 1), 1 μL of 1 U Top DNA polymerase (Bioneer), 2 μL of cDNA, and 11 μL of DEPC-treated water.
3. The semi-nested PCR (snPCR) was performed in a 20 μL reaction mixture containing 2 μL of 10 × PCR buffer (Bioneer), 2 μL of 10 mM dNTP (Bioneer), 1 μL of 20 pM GI-F2 and GI-R1 primers for HuNoV GI or 1 μL of 20 pM GII-F2 and GII-R1 primers for HuNoV GII (Table 1), 1 μL of 1 U

Top DNA polymerase (Bioneer), 2 μL of the first PCR product, and 11 μL of DEPC-treated water.

4. PCR was performed on the T100 thermal cycler (Bio-Rad) with the following program: initial denaturation at 94 °C for 5 min; 35 cycles of denaturation at 94 °C for 30 s, annealing at 55 °C for 30 s, and extension at 72 °C for 90 s; followed by a final extension at 72 °C for 7 min.
5. The final PCR products were visualized by electrophoresis on a 1.5% agarose gel. Amplicons for HuNoV GI and GII were identified as 313 bp and 310 bp, respectively, confirming their expected sizes.

3.7 Real-Time RT-PCR

1. Real-time RT-PCR was performed in a total volume of 25 μL, which included 12.5 μL of 2 × Premix Ex Taq (TaKaRa Bio Inc.), 0.5 μL of 10 pM COG1F and COG1R primers, 0.5 μL of 15 pM RING1(a), and 1.5 μL of 5 pM RING1(b) probes for HuNoV GI [8]. For HuNoV GII, the reaction contained 0.5 μL of 10 pM COG2F and COG2R primers and 2 μL of 5 pM RING2 probes (Table 1) [8]. Additionally, 2.5 μL of cDNA and 7 μL of DEPC-treated water were included to reach the final volume.
2. Real-time RT-PCR was performed on a CFX96 Touch Real-Time PCR Detection System (Bio-Rad). The cycling conditions consisted of an initial denaturation step at 95 °C for 10 min, followed by 45 cycles of amplification, each including denaturation at 95 °C for 15 s and annealing/extension at 56 °C for 60 s.
3. Fluorescence signals were measured at the end of each annealing/extension step.
4. Cycle threshold values were plotted against log-transformed RNA concentrations to calculate assay efficiency.

3.8 Specificity Testing

1. To evaluate assay specificity, nucleic acids were extracted from HAV, HEV, rotavirus, and adenovirus.
2. Positive controls included HuNoVs GI and GII RNAs confirmed by sequence analysis.
3. To confirm the specificity of the assay, RFLP analysis was performed. *Hha*I and *Hae*III restriction enzymes were used to digest HuNoV GI and GII LAMP products, respectively, following the manufacturer's protocol (Bioneer) (Fig. 2).

3.9 Sensitivity Testing

1. The sensitivity of two types of RT-LAMP, snRT-PCR and real-time RT-PCR assays, was determined using tenfold serial

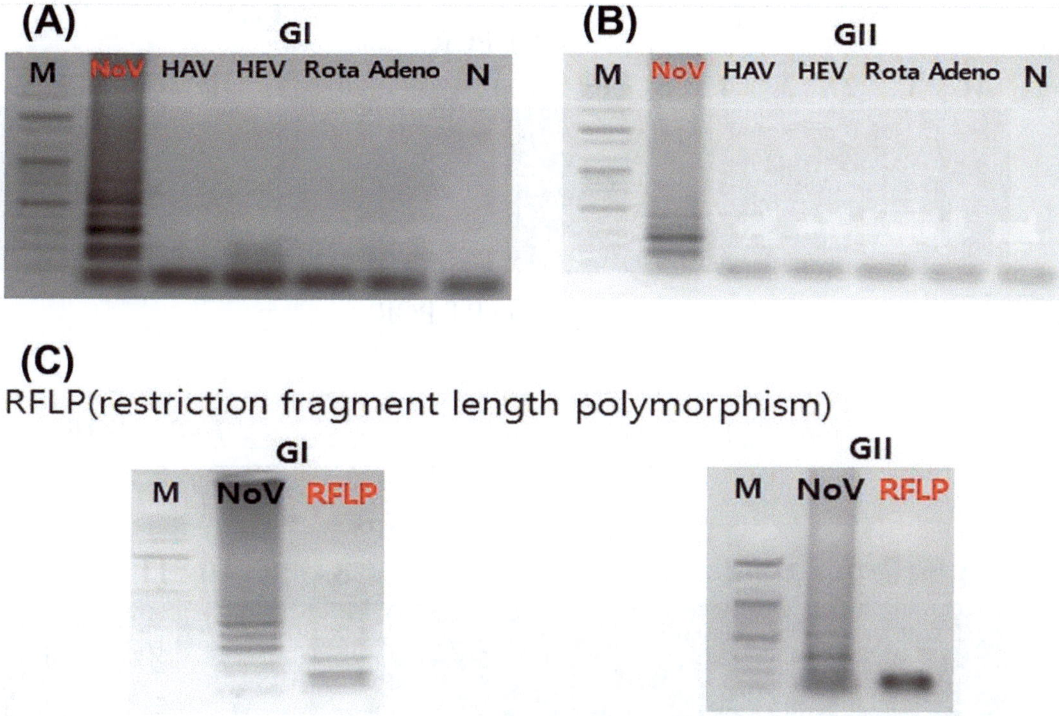

Fig. 2 Specificity test of one-step RT-LAMP. (**a**) HuNoV GI and (**b**) HuNoV GII. Lane M, 100-bp DNA marker; HuNoV GI.1, HuNoV GII.4, HAV, HEV, rotavirus (Rota), adenovirus 41 (Adeno), negative control (N). (**c**) HuNoV GI LAMP product digested with *Hha*I, HuNoV GII LAMP product digested with *Hae*III

dilutions of ATCC synthetic standard RNA for HuNoV GI and GII.

2. RNA concentrations ranged from 10^0 to 10^6 copies/μL.
3. The limit of detection was defined as the lowest RNA concentration that consistently produced a positive result in three independent replicates (Fig. 3).

3.10 Viral Recovery from Oysters

1. Oysters were artificially contaminated with HuNoVs GI and GII by immersing them in seawater spiked with 10^6 copies/mL of synthetic standard RNA for 24 h at 15 °C.
2. After contamination, the oysters were rinsed with sterile seawater, and 2 g of digestive tissues were processed as described in the "Sample Collection and Preparation" section.
3. Viral RNA was extracted, and detection was performed using both one-step RT-LAMP, real-time RT-PCR (Table 2).

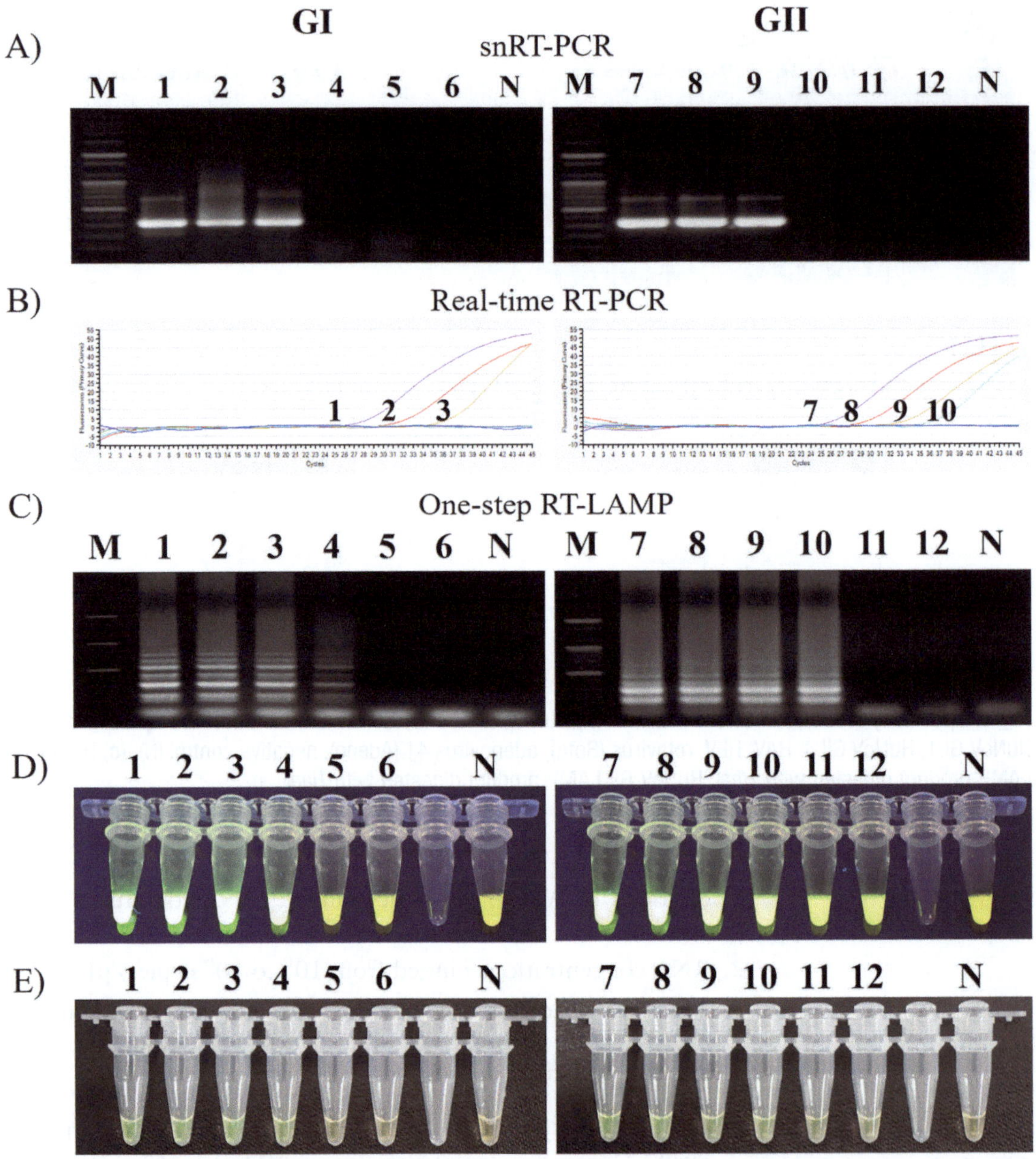

Fig. 3 Sensitivity comparison among snRT-PCR, real-time RT-PCR, and one-step RT-LAMP for detecting HuNoV GI and GII. (**a**) snRT-PCR showing final amplifications, (**b**) real-time RT-PCR, and (**c**) one-step RT-LAMP. Lane M, 100-bp DNA marker; lane 1, 4.7×10^4 HuNoV genome copies (GC)/mL; lane 2, 4.7×10^3 HuNoV GC/mL; lane 3, 4.7×10^2 HuNoV GC/mL; lane 4, 4.7×10^1 HuNoV GC/mL; lane 5, 4.7×10^0 HuNoV GC/mL; lane 6, 4.7×10^{-1} HuNoV GC/mL; lane 7, 4.8×10^4 HuNoV GC/mL; lane 8, 4.8×10^3 HuNoV GC/mL; lane 9, 4.8×10^2 HuNoV GC/mL; lane 10, 4.8×10^1 HuNoV GC/mL; lane 11, 4.8×10^0 HuNoV GC/mL; lane 12, 4.8×10^{-1} HuNoV GC/mL; lane N, negative control. (Permitted from Jeon et al. [11])

Table 2
Comparison of snRT-PCR, real-time RT-PCR, real-time RT-LAMP, and one-step RT-LAMP for the detection of HuNoV

HuNoV detection		Limit of detection	Stool	Hemocytes	Digestive tissue
GI	snRT-PCR	10^2	4/4	5/6	4/6
	Real-time RT-PCR	10^2	4/4	6/6	6/6
	Real-time RT-LAMP	10^5	4/4	1/6	ND[a]
	One-step RT-LAMP	10^1	4/4	6/6	6/6
GII	snRT-PCR	10^2	6/6	4/6	1/6
	Real-time RT-PCR	10^1	6/6	6/6	2/6
	Real-time RT-LAMP	Not detected	0/6	0/6	ND
	One-step RT-LAMP	10^1	6/6	6/6	1/6

ND not done

References

1. Hall AJ, Glass RI, Parashar UD (2016) New insights into the global burden of noroviruses and opportunities for prevention. Expert Rev Vaccines 15(8):949–951
2. Omatola CA, Mshelbwala PP, Okolo M-LO, Onoja AB, Abraham JO, Adaji DM et al (2024) Noroviruses: evolutionary dynamics, epidemiology, pathogenesis, and vaccine advances – a. Comprehen Rev Vacc 12(6):590
3. Tarek F, Hassou N, Benchekroun MN, Boughribil S, Hafid J, Sail R et al (2020) Emergence of norovirus strains in worldwide distribution and environmental spreading. In: Emerging and reemerging viral pathogens. Elsevier, pp 981–998
4. Winder N, Gohar S, Muthana M (2022) Norovirus: an overview of virology and preventative measures. Viruses 14(12):2811
5. Sun Y, Liang M, Zhao F, Su L (2023) Research progress on biological accumulation, detection and inactivation technologies of norovirus in oysters. Food Secur 12(21):3891
6. Pouillot R, Smith M, Van Doren JM, Catford A, Holtzman J, Calci KR et al (2022) Risk assessment of norovirus illness from consumption of raw oysters in the United States and in Canada. Risk Anal 42(2):344–369
7. de Bruin E, Duizer E, Vennema H, Koopmans MP (2006) Diagnosis of norovirus outbreaks by commercial ELISA or RT-PCR. J Virol Methods 137(2):259–264
8. Kim S-H, Cheon D-S, Kim J-H, Lee D-H, Jheong W-h, Heo Y-J et al (2005) Outbreaks of gastroenteritis that occurred during school excursions in Korea were associated with several waterborne strains of norovirus. J Clin Microbiol 43(9):4836–4839
9. Kageyama T, Kojima S, Shinohara M, Uchida K, Fukushi S, Hoshino FB et al (2003) Broadly reactive and highly sensitive assay for Norwalk-like viruses based on real-time quantitative reverse transcription-PCR. J Clin Microbiol 41(4):1548–1557
10. Fukuda S, Takao S, Kuwayama M, Shimazu Y, Miyazaki K (2006) Rapid detection of norovirus from fecal specimens by real-time reverse transcription-Loop-mediated isothermal amplification assay. J Clin Microbiol 44(4): 1376–1381
11. Jeon SB, Seo DJ, Oh H, Kingsley DH, Choi C (2017) Development of one-step reverse transcription Loop-mediated isothermal amplification for norovirus detection in oysters. Food Control 73:1002–1009
12. Provost K, Dancho BA, Ozbay G, Anderson RS, Richards GP, Kingsley DH (2011) Hemocytes are sites of enteric virus persistence within oysters. Appl Environ Microbiol 78(20): 7455–7466

Chapter 11

Hepatitis A Virus Detection Using the Loop-Mediated Isothermal Amplification-Based (LAMP) Technique in Foods

Rafael Dorighello Cadamuro, Mariana Alves Elois, Nadine Yeramian, Daniel Pérez-Alonso, Gislaine Fongaro, and David Rodríguez-Lázaro

Abstract

This protocol describes a rapid and sensitive method for detecting hepatitis A virus (HAV) in food samples using reverse transcription loop-mediated isothermal amplification (RT-LAMP). The technique targets the conserved 5′-UTR region of the HAV genome with six specific primers, enabling simultaneous reverse transcription and amplification at 62.5 °C for 60 min. This method offers a quick and reliable approach for HAV detection in food, supporting food safety monitoring and outbreak investigations.

Key words Hepatitis A virus (HAV), RT-LAMP, Food detection, Isothermal amplification, Real-time monitoring

1 Introduction

Hepatitis A virus (HAV) is one of the five hepatotropic viruses that cause liver inflammation. It is transmitted through the ingestion of food or water contaminated with feces or by contact with an infected person. HAV is a 27 nm, non-enveloped RNA virus belonging to the Picornaviridae family. The World Health Organization estimates that HAV causes over 100 million infections per year [1]. With developing countries being the most affected by water contamination, there is an urgent need to detect this virus at a relatively low cost, and the LAMP technique is a promising alternative. Numerous studies have demonstrated the detection of HAV using the LAMP method, as it can efficiently amplify RNA sequences through combination with reverse transcription [2, 3].

Nucleic acid amplification is a valuable tool for detecting pathogens, and the invention of PCR by Kary Mullis in 1983 was a major revolution in molecular biology. Since then, this method, used in many scientific fields, has continued to evolve and be optimized. A newly developed gene amplification technique, known as LAMP,

Alejandro Garrido-Maestu and Alexandre Lamas (eds.), *Foodborne Pathogens: Loop-mediated Isothermal Amplification Protocols*, Methods and Protocols in Food Science, https://doi.org/10.1007/978-1-0716-4526-0_11,

has emerged to improve certain aspects of PCR. It is characterized by its speed, cost-effectiveness, and high specificity. Developed by Notomi in 2000, it can amplify a few copies of DNA to 10^9 in less than an hour under isothermal conditions, with great specificity [4].

The loop-mediated isothermal amplification (LAMP) method is a technique that facilitates the rapid amplification of DNA through auto cycling strand displacement synthesis. This process is driven by a DNA polymerase exhibiting high strand displacement activity, in conjunction with a carefully designed set of primers. The LAMP reaction involves two pairs of essential primers: the forward inner primer (FIP) and the backward inner primer (BIP), as well as the forward outer primer (F3) and the backward outer primer (B3). Additionally, using two extra primers—loop forward (LF) and loop backward (LB)—is recommended, as they can reduce the amplification time from 1 h to 30 min [4].

In the early phase of the LAMP reaction, all four primers are incorporated; however, as the cyclic amplification progresses, only the inner primers are employed for strand displacement synthesis. Each inner primer consists of two distinct sequences that correspond to the sense and antisense strands of the target DNA, enabling priming during the initial stage and facilitating self-priming in subsequent cycles.

The amplification process begins with the formation of a dumbbell-like structure characterized by stem-loop formations at each end. This structure serves as the foundational template for further amplification cycles. As the reaction continues, amplification occurs in a cyclical manner, with each newly synthesized strand displacing the previous one during elongation. The end product is a mixture of stem-loop DNA with cauliflower-like structures [4], as presented in Fig. 1.

This method presents advantages compared to the PCR technique:

- The entire reaction occurs at a constant temperature, avoiding the time loss associated with temperature changes. As a result, the amount of amplified DNA can reach up to a billion copies, compared to the millions produced by PCR [5].
- Since there are no heating and cooling cycles, a thermocycler is not needed. Instead, the reaction can occur in a dry block heater or water bath, reducing costs [4].
- The amplification process is faster than PCR, particularly when using the two additional loop primers [3].
- LAMP can achieve sensitivity up to 1.000 times higher than conventional or multiplex PCR [6].
- The technique can be performed in the field and does not require highly qualified personnel.

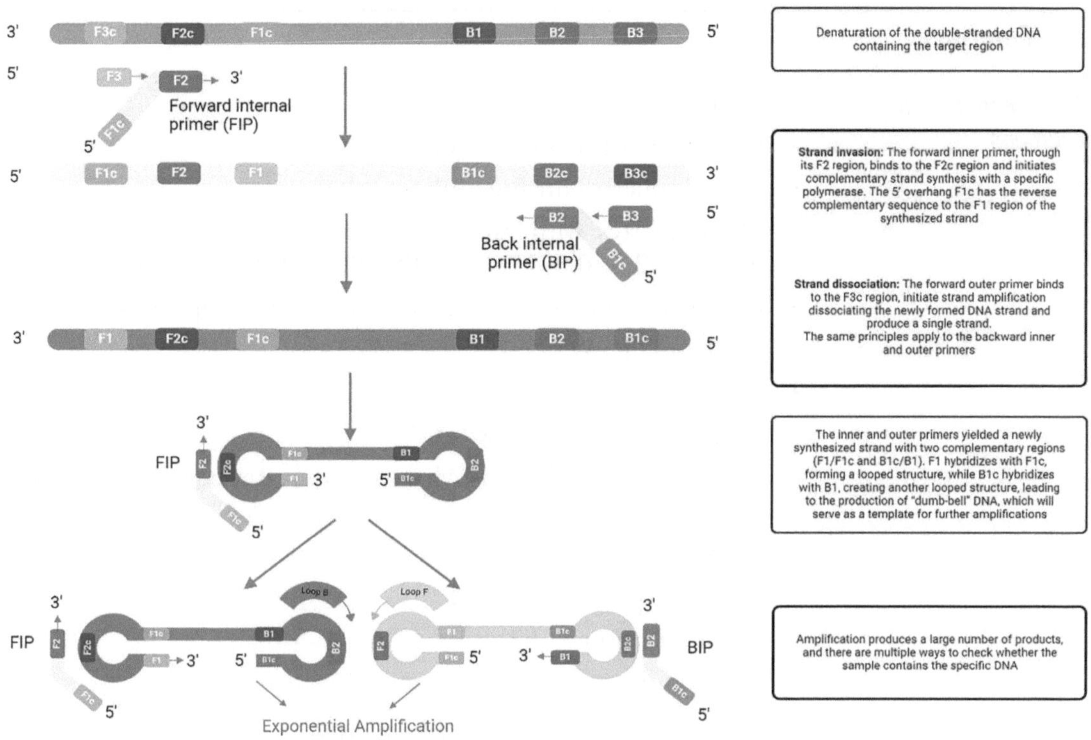

Fig. 1 Steps of process of amplification and detection of region target

- The LAMP technique can be visualized by eye and does not require sophisticated devices.
- Thanks to the use of two to three pairs of primers, this technique is highly specific. It can identify up to eight regions on the target gene, compared to only two in PCR [7].

However, LAMP does have some limitations, such as:

- Primer design is relatively complex because it requires the design of four to six specific primers.
- It appears to be sensitive to cross-contamination, such as material present in aerosols [7].
- The use of multiple primers increases the risk of primer–primer hybridizations resulting in template-free amplification, leading to false-positive results [8].

Significant work is being done to overcome these limitations.

Due to its simplicity, high specificity, speed, and low cost, the LAMP technique has been used to identify pathogens in plants, animals, and humans [7], including bacteria, viruses, parasites, and allergens [8]. This chapter aims to present a protocol for the identification of hepatitis A virus in food samples.

2 Materials

2.1 Reagents and Solutions

- Cooler or container with ice.
- Sterile microtubes of 1.5 mL.
- 0.2 mL microcentrifuge tubes
- Micropipettes with variable volumes and their corresponding tips (0.1–2.5; 0.5–10; 10–100; 100–1000 μL; 1–10 mL).
- Gel casting tray using tape, combs, and electrophoresis system.
- UV lamp (15 W).
- Enzymes:
  - *Bst* DNA Polymerase: 16 U.
  - Avian myeloblastosis virus (AMV) Reverse Transcriptase: 2 U.
- Primers:
  - Inner primers: FIP, BIP.
  - Outer primers: F3, B3.
  - Loop primers: FL, BL.
- 2× reaction mixture: 12.5 μL
- Extracted RNA: 5 μL.
- Nuclease-free water.
- Restriction enzyme AflII.
- Agarose powder.
- Ethidium bromide.
- Loading buffer.
- Molecular weight marker (100-bp DNA ladder).

2.2 Equipment

- Loopamp Real-Time Turbidimeter: LA-320C, Teramecs, Kyoto, Japan.
- UV transilluminator: For visualization of products on agarose gel.
- Agarose Gel Electrophoresis System: For analysis of amplified products.
- PCR thermocycler.
- Heat block or water bath.
- Vortex.
- Minicentrifuge.
- Laminar flow hoods or PCR cabinet.
- Water purification system (Milli-Q, Millipore).
- Microwave.

3 Methods

3.1 Primer Design

1. Design primers based on the highly conserved 5′-UTR of the HAV genome using PrimerExplorer V3 (*see* **Note 1**).
2. Use sequences from the KRM238 strain (genotype IIIB) for the primer template (*see* **Note 2**).
3. Include a set of six primers: two outer (F3 and B3), two inner (FIP and BIP), and two loop primers (F loop and B loop) (*see* **Note 3**).

3.2 RT-LAMP Assay Procedure

1. Prepare a PCR master mix in 1.5 mL microtubes according to the specifications in Table 1. The final volume of the master reaction depends on the reaction number required for each PCR. Therefore, prepare 10% excess of the master mix (*see* **Note 4**).

1. Set up the PCR in individual 0.2 mL microtubes, according to the sample number, including nontemplate (NTC), negative controls, positive controls, and the test samples, each at least in duplicate (*see* **Note 5**).
2. Mix and centrifuge (spin) the master mix and dispense equal aliquots into each PCR tube.
3. Add the RNA template (5 μL of RNA per sample to each well) to each reaction tube (remember that the negative control does not receive any template).
4. Centrifuge the reaction tubes at low speed for at least 1 min and load the plate/tubes onto a heat block or a water bath.

Table 1
Relation of components of reactions of LAMP and concentrations

Component	Stock concentration	Amount per 25 μL reaction (μL)	Final concentration
2× reaction mixture	2×	12.5	1×
FIP primer	100 μM	2	0.8 μM
BIP primer	100 μM	4	1.6 μM
F3 primer	100 μM	0.5	0.2 μM
B3 primer	100 μM	0.5	0.2 μM
Loop F primer	100 μM	2	0.8 μM
Loop B primer	100 μM	2	0.8 μM
Bst DNA polymerase	8000 U/mL	2	640 U/mL
AMV reverse transcriptase	10,000 U/mL	0.2	80 U/mL
Template RNA	Variable	5	Variable
Nuclease-free water	–	To 25	–

3.3 Amplification

1. Incubate the reaction mixtures at 62.5 °C for 60 min in a Loopamp real-time turbidimeter (LA-320C, Teramecs, Kyoto, Japan) (*see* **Note 6**).
2. Perform real-time monitoring every 6 s by spectrophotometric analysis, recording the optical density (OD) at 650 nm (*see* **Note 7**).
3. Include a negative control using water, which will be analyzed for 60 min to establish a turbidity signal threshold for distinguishing positive from negative results. A turbidity signal greater than 0.1 within 50 min indicates a positive result (*see* **Note 8**).

3.4 Enzymatic Digestion

1. Validate the specificity of the RT-LAMP amplified products from HAV strain KRM238 by performing restriction digestion with the AflII enzyme after purifying the RT-LAMP products using a column-based PCR Purification Kit (*see* **Note 9**).

3.5 Electrophoresis Gel

1. Prepare a 2% agarose gel by weighing and adding 2 g of agarose powder to a 200 mL Erlenmeyer flask.
2. Add 100 mL of TAE 1× buffer.
3. Boil in a microwave for 15–20 s, remove the flask, and mix gently a couple of times to dissolve homogeneously, repeating the heating up to three times.
4. Let the agarose cool until the recipient can be touched.
5. Add ethidium bromide solution to the gel and mix gently so bubbles do not form.
6. Seal both the open ends of the gel casting tray using tape, insert the combs, and pour the melted agarose.
7. Once the gel hardens, remove the tapes and withdraw the combs gently without tearing the wells.
8. Place the gel casting tray containing the set gel in the electrophoresis tank and add a sufficient volume of electrophoretic buffer (TAE 1×) to cover the gel up to 1 mm deep. Avoid air pocket formation within the wells.

3.6 Load and Run

1. Add 5 μL of loading buffer to each reaction tube removed from the heat block/water bath.
2. Load the first well with 5 μL of the molecular weight marker (100-bp DNA ladder).
3. Load the successive wells with 10 μL of each sample, NTC, and positive control.
4. Attach the leads according to their specific colors and set the voltage to 120 V for 30–45 min to run the gel (normally 1–10 V/cm of gel).

3.7 UV Light Transilluminator and Bands Visualization

1. To view the bands, place the gel on the UV light transilluminator.
2. Operate the software and equipment according to the manufacturer's recommendations.
3. Analyze both the digested and non-digested products on the 2% agarose gel, visualizing the bands under the ultraviolet transilluminator (*see* **Note 10**).
4. Incubate the reaction mixtures at 62.5 °C for 60 min in a Loopamp real-time turbidimeter.
5. Perform real-time monitoring every 6 s by spectrophotometric analysis, recording the optical density (OD) at 650 nm.
6. Include a negative control using water, which will be analyzed for 60 min to establish a turbidity signal threshold for distinguishing positive from negative results. A turbidity signal greater than 0.1 within 50 min indicates a positive result.
7. Validate the specificity of the RT-LAMP amplified products from HAV strain KRM238 by performing restriction digestion with the AflII enzyme after purifying the RT-LAMP products using a column-based PCR Purification Kit.
8. Analyze both the digested and non-digested products on a 2% agarose gel, staining the bands with ethidium bromide and visualizing them under an ultraviolet transilluminator.

4 Notes

1. Primers should be designed using primer design software, targeting the highly conserved 5′-UTR region of the HAV genome. Synthesized LAMP primers should be stored at − 20 °C at 100 μM in stock aliquots.
2. The KRM238 strain (genotype IIIB) sequences are recommended for use as the primer template due to their high conservation among HAV strains.
3. The set of six primers (two outer, two inner, and two loop primers) is essential for efficient LAMP amplification. Ensure all primers are included in the reaction mix.
4. When preparing the master mix, calculate for a 10% excess to account for pipetting errors. For a large number of samples, 5% excess may be sufficient, while for experienced technicians, a 2% excess can be adequate.
5. Always include appropriate controls in your experimental setup. Negative control (using water instead of template) is crucial for establishing the turbidity threshold and ensuring the absence of contamination.

6. The incubation temperature (62.5 °C) and time (60 min) are optimized for HAV detection. Ensure the Loopamp real-time turbidimeter is properly calibrated before use.
7. Real-time monitoring every 6 s at 650 nm allows for precise detection of amplification. A turbidity signal greater than 0.1 within 50 min indicates a positive result.
8. The negative control analysis for 60 min is crucial for establishing the turbidity signal threshold. This helps in distinguishing true positive results from false positives.
9. Restriction digestion with AflII enzyme is used to validate the specificity of the RT-LAMP amplified products. This step helps confirm that the amplified sequences correspond to the HAV target.
10. When analyzing the gel, both digested and non-digested products should be included. This allows for comparison and confirmation of specific amplification.

References

1. Cao G, Jing W, Liu J, Liu M (2021) The global trends and regional differences in incidence and mortality of hepatitis a from 1990 to 2019 and implications for its prevention. Hepatol Int 15(5):1068–1082. https://doi.org/10.1007/S12072-021-10232-4
2. Wu R, Meng B, Corredig M, Griffiths MW (2023) Rapid detection of hepatitis a virus in foods using a bioluminescent assay in real-time (BART) and reverse transcription Loop-mediated isothermal amplification (RT-LAMP) technology. Food Environ Virol 15(2): 144–157. https://doi.org/10.1007/S12560-022-09548-7/TABLES/3
3. Yoneyama T, Kiyohara T, Shimasaki N, Kobayashi G, Ota Y, Notomi T, Totsuka A, Wakita T (2007) Rapid and real-time detection of hepatitis a virus by reverse transcription loop-mediated isothermal amplification assay. J Virol Methods 145(2):162–168. https://doi.org/10.1016/J.JVIROMET.2007.05.023
4. Notomi T, Okayama H, Masubuchi H, Yonekawa T, Watanabe K, Amino N, Hase T (2000) Loop-mediated isothermal amplification of DNA. Nucleic Acids Res 28(12):e63–e63. https://doi.org/10.1093/NAR/28.12.E63
5. Ushikubo H (2004) LAMP法の原理 —遺伝子の簡易・迅速な増幅法—. ウイルス 54(1): 107–112. https://doi.org/10.2222/JSV.54.107
6. Wang X, Seo DJ, Lee MH, Choi C (2014) Comparison of conventional PCR, multiplex PCR, and Loop-mediated isothermal amplification assays for rapid detection of *Arcobacter* species. J Clin Microbiol 52(2):557–563. https://doi.org/10.1128/JCM.02883-13/ASSET/E5E7177F-6EA1-4680-BEB7-6397CCF3A973/ASSETS/GRAPHIC/ZJM9990931570003.JPEG
7. Soroka M, Wasowicz B, Rymaszewska A (2021) Loop-mediated isothermal amplification (LAMP): the better sibling of PCR? Cells 10: 1931. https://doi.org/10.3390/CELLS10081931
8. Garg N, Ahmad FJ, Kar S (2022) Recent advances in Loop-mediated isothermal amplification (LAMP) for rapid and efficient detection of pathogens. Curr Res Microb Sci 3:100120. https://doi.org/10.1016/J.CRMICR.2022.100120

Chapter 12

Hepatitis E Virus Detection Using the Loop-Mediated Isothermal Amplification-Based (LAMP) Technique

Mariana Alves Elois, Rafael Dorighello Cadamuro, Lorena Casado Martin, Jorge Santamaría Palacios, Gislaine Fongaro, and David Rodríguez-Lázaro

Abstract

Loop-mediated isothermal amplification (LAMP) allows for the rapid, efficient, and cost-effective amplification of nucleic acids. In this chapter, we propose an RT-LAMP protocol for the detection of hepatitis E virus (HEV), an enterovirus of public health significance. The process begins with the preparation of a master mix containing the appropriate primers. The reaction is then carried out at a constant temperature of 63 °C for 45 min. During this step, RNA is reverse transcribed into cDNA, which is subsequently amplified by specific enzymes. Following amplification, an enzymatic digestion is performed, and the resulting products are detected via electrophoresis. Alternatively, other detection methods can be employed, such as monitoring the accumulation of magnesium pyrophosphate, observing visual color changes, detecting fluorescent signals, or using real-time detection techniques.

Key words RT-LAMP, Hepatitis E virus, Detection, Quantification, Protocol

1 Introduction

Loop-mediated isothermal amplification (LAMP) is a molecular technique known for its specificity, efficiency, and rapidity in DNA amplification. Initially described by Notomi et al. [1], this technique has evolved, incorporating other molecular approaches such as reverse transcription and multiplex amplification. Various detection methods and formats, including electric [2] and in-disc LAMP [3] techniques, have also been developed. This molecular method enables the detection of infectious diseases caused by microorganisms in humans, livestock, and plants [4]. A wide range of pathogens can be identified using LAMP, including significant bacteria such as *Streptococcus* spp., *Klebsiella pneumoniae* [5, 6], *Salmonella typhi* [7], and *Listeria monocytogenes* [8]; protozoa like *Plasmodium* spp. [9–11], and, with some additional effort, even fungi.

Alejandro Garrido-Maestu and Alexandre Lamas (eds.), *Foodborne Pathogens: Loop-mediated Isothermal Amplification Protocols*, Methods and Protocols in Food Science, https://doi.org/10.1007/978-1-0716-4526-0_12,

Moreover, LAMP is effective in detecting a variety of viruses, including adenovirus [12], varicella-zoster virus [13], cytomegalovirus [14], and hepatoviruses such as hepatitis B [15], C [16], or E.

When RNA viruses are the target of study, reverse transcription LAMP (RT-LAMP) is required. Similar to RT-PCR, RT-LAMP includes a reverse transcription step to generate complementary DNA from RNA, facilitated by reverse transcriptase, which is subsequently amplified by a DNA polymerase. This process can be performed in a single step, allowing incubation at a constant temperature with all necessary primers and enzymes [1, 17]. This simplicity is one of the reasons LAMP is proposed as a valuable tool for epidemiological surveillance [4].

Enteroviruses are responsible for several diseases in humans, and traditional methods for their detection, such as cell culture assays, neutralizing antisera, RT-PCR, or RT-qPCR, are often more expensive, complex, and time-consuming [18]. This makes loop-mediated isothermal amplification (LAMP) an attractive alternative. Within the enterovirus group, the hepatitis E virus (HEV) stands out as an enteric pathogen of humans and animals, with pigs being its primary reservoir, making it a zoonotic disease [19]. HEV is responsible for epidemics and endemics of acute hepatitis in humans [20] posing a significant public health threat. The virus is typically transmitted via the fecal-oral route, either between animals, from animals to humans, or through contaminated food or water sources [21].

Given the public health significance of HEV in both developed and developing countries, a rapid and effective detection method is crucial, particularly for the latter. As a result, various RT-LAMP detection protocols have been proposed [22–25], and this chapter suggests a potential method for detecting this pathogen.

2 Materials

2.1 Materials

1. Cooler or container with ice.
2. Sterile microtubes of 1.5 mL.
3. 0.2 mL microcentrifuge tubes
4. Erlenmeyer flask.
5. Tape.
6. Micropipettes of variable volume and their respective tips (0.1–2.5; 0.5–10; 10–100; 100–1000 μL; 1–10 mL).
7. Gel casting tray using tape, combs, and electrophoresis system.
8. UV lamp (15 W).

2.2 Equipment

1. Heat block or water bath.
2. Vortex.
3. Minicentrifuge.
4. Laminar flow hoods or PCR cabinet.
5. Water purification system (Milli-Q, Millipore).
6. Microwave.
7. UV transilluminator.
8. Spectrophotometer, real-time turbidity meter LA-200, or similar devices.

2.3 Reagents

1. TAE buffer (10×): Prepare 48.4 g of Tris base, 11.42 mL of glacial acetic acid, 20 mL of 0.5 M EDTA (pH 8.0), and dissolve in 1000 mL of water. Dilute 1: 10 with water to make 1× TAE.
2. Oligonucleotide primers (Table 1) (*see* **Note 1**).
3. dNTPs (10 mM).
4. $(NH_4)_2SO_4$ (100 mM).
5. $MgSO_4$ (100 mM) (*see* **Note 2**).

Table 1
Primers are used in this protocol for HEV detection in food, as described herein

Primer description	Primer name	Sequence (5′–3′)	Genome position	Concentration	References
Outer primer	F3	TGGAGATGCCACCATGCGCTC TCG	5175–5198	Final volume: 25 μL 2 μL of RNA 5 pmol	[23, 26]
Outer primer	B3	ACGCCAAGCGGAGCCGAG TGGAC	5445–5423	5 pmol	[23, 26]
Inner primer	FIP	ACCGCCGCTGCGCCGCCCC TGCTCTTCGTGCTTCTGCCTA TGCTGC	5296–5279 5213–5240	40 pmol	[26]
Inner primer	BIP	GGTGGTTTCTGGGG TGACCGGGTTCCGAAGGGG TTGGTTGGATGAATATAGG	5303–5326 5376–5349	40 pmol	[26]
Loop primer	LOOP-F	CGACGGCCAGACGGCTGACC	5273–5254	20 pmol	[26]
Loop primer	LOOP-B	GATTCTCAGCCCTTCGCCCTCC	5320–5342	20 pmol	[26]

6. KCl (100 mM).
7. Tris-HCl (pH 8.8) (100 mM).
8. *Bst* DNA polymerase (8 U) (*see* **Note 3**).
9. PrimeScript reverse transcriptase (200 U) (*see* **Note 4**).
10. Molecular biology grade water.
11. Restriction enzyme BsiEI.
12. Agarose powder.
13. GreenSafe.
14. Loading buffer.
15. Molecular weight marker (100-bp DNA ladder).
16. Loopamp fluorescent detection reagent.
17. SYBR Green I dye.

3 Methods

3.1 RNA Extraction

1. Viral RNA was extracted with a viral RNA extraction kit, according to the manufacturer's instructions (*see* **Note 5**).

3.2 RT-Lamp

Precautions must be taken to avoid contamination of samples and reagents during the preparation of PCR reactions (*see* **Note 6**).

1. Arrange all reagents needed for PCR in a cooler or container with ice and let them thaw completely before starting the procedure (*see* **Note 7**).
2. After the reagents have been completely thawed, prepare a PCR master mix in 1.5 mL microtubes according to the specifications in Table 2. The final volume of the master reaction depends on the reaction number required for each PCR. Therefore, prepare 10% excess of the master mix (*see* **Note 8**).
3. Set up the PCR in individual 0.2 mL microtubes, according to the sample number, including no template control (NTC) (*see* **Note 9**), negative controls, positive controls, and the test samples, each at least in duplicate (*see* **Note 10**).
4. Mix and centrifuge (spin) the master mix and dispense equal aliquots into each PCR tube.
5. Add the RNA template (2 μL of RNA per sample to each well) to each reaction tube (remember that the negative control does not receive any template).
6. Centrifuge the reaction tubes at low speed for at least 1 min and load the plate/tubes onto a heat block or a water bath.

Table 2
Composition of PCR reactions

Mix components	Stock concentration	Final concentration	Mix 1 (μL)	Mix N
F3	100 μM (100 pmol/μL)	5 pmol	0.05	0.05 μL × N μL
B3	100 μM (100 pmol/μL)	5 pmol	0.05	0.05 μL × N μL
FIP	100 μM (100 pmol/μL)	40 pmol	0.4	0.4 μL × N μL
BIP	100 μM (100 pmol/μL)	40 pmol	0.4	0.4 μL × N μL
LOOP-F	100 μM (100 pmol/μL)	20 pmol	0.2	0.2 μL × N μL
LOOP-B	100 μM (100 pmol/μL)	20 pmol	0.2	0.2 μL × N μL
dNTPs	10 mM	2.8 mM	7	7 μL × N μL
$(NH_4)_2SO_4$	100 mM	10 mM	2.5	2.5 μL × N μL
$MgSO_4$	100 mM	12 mM	3	3 μL × N μL
KCl	100 mM	10 mM	2.5	2.5 μL × N μL
Tris-HCl (pH 8.8)	100 mM	20 mM	5	5 μL × N μL
Bst DNA polymerase	8 U	8 U	1	1 μL × N μL
PrimeScript reverse transcriptase	200 U	50 U	0.25	0,25 μL × N μL
Molecular biology grade water	q.s.	q.s.	0.45	0,45 μL × N μL

Table 3
The PCR profile of the PCR system is described herein

Step	Temperature (°C)	Time (s)
Amplification	63	45
Inactivation	80	5

3.3 Amplification

1. Operate the equipment (heat block or water bath) according to the manufacturer's recommendations and the experiment profile described in Table 3.

3.4 Enzymatic Digestion

1. Add 1 μL of the amplified products in a 20 μL total reaction volume with the restriction enzyme BsiEI for 2 h at 60 °C (*see* **Note 11**).

3.5 Electrophoresis Gel

1. Weigh and add 1.5 g of agarose powder to a 200 mL Erlenmeyer flask.
2. Add 100 mL of TAE 1× buffer.
3. Boil in a microwave for 15–20 s, remove the flask, and mix gently a couple of times to dissolve homogeneously, repeating the heating up to three times (*see* **Note 12**).
4. Let the agarose cool until 35–40 °C.
5. Add 3 μL of GreenSafe solution to each 100 mL of gel and mix gently so bubbles do not form (*see* **Note 13**).
6. Seal both the open ends of the gel casting tray using tape, insert the combs, and pour the melted agarose (*see* **Note 14**).
7. Once the gel hardens, remove the tapes, and withdraw the combs gently without tearing the wells.
8. Place the gel casting tray containing the set gel in the electrophoresis tank and add a sufficient volume of electrophoretic buffer (TAE 1×) to cover the gel up to 1 mm deep. Avoid air pocket formation within the wells (*see* **Note 15**).

3.6 Load and Run

1. Add 10 μL of loading buffer to each reaction tube removed from the heat block/water bath (*see* **Note 16**).
2. Load the first well with 5 μL of the molecular weight marker (100-bp DNA ladder).
3. Load the successive wells with 10 μL of each sample, NTC, and positive control (*see* **Note 17**).
4. Attach the leads according to their specific colors and set the voltage to 120 V for 30–45 min to run the gel (normally 1–10 V/cm of gel) (*see* **Note 18**).
5. Switch off the electricity supply when the dye from the loading buffer has moved a distance sufficient for the DNA separation.

3.7 UV Light Transilluminator and Bands Visualization

1. To view the bands, place the gel on the UV light transilluminator (*see* **Note 19**).
2. Operate the software and equipment according to the manufacturer's recommendations.
3. Briefly, turn on the computer connected to the photo documentation software and the photo documenter. Turn on the photo documenter. Place the gel on the UV light protector screen and close it. Initialize the software, select the equipment, and choose *Auto Expose*. Observe the image and adjust the exposure until it is properly displayed (freeze). Once the desired image is visible, save it by selecting *Save* and *Export*.

3.8 Other RT-LAMP Product Detection Methods

3.8.1 Accumulation of Magnesium Pyrophosphate

1. Prepare the reaction mix using the components shown in Table 2, and follow Subheading 3.1 until **step 5**.
2. Place the reaction tubes in the hot plate of real-time turbidity meter LA-200 or a similar device at 63 °C for 60 min and inactivate at 80 °C for 5 min. The RT-LAMP products will be monitored and quantified by measuring the accumulation of magnesium pyrophosphate at 650 nm with a threshold equal to or higher than 0.2 within 45 min [24, 27].

3.8.2 Analysis of RT-LAMP Reactions by a Visual Color Change and Fluorescent Signal

1. Prepare the reaction mixture using the components shown in Table 2. However, add 2 μL SYBR Green I dye or 1 μL Loopamp fluorescent detection reagent per 25 μL reaction, and reduce the volume of water in the reaction mixture accordingly.
2. Follow the **steps 3–6** of the Subheading 3.1.
3. Observe reaction products under normal or UV light conditions (wavelength at 240 ~ 260 nm or 350 ~ 370 nm) (*see* **Note 20**). Fluorescence spectrophotometry of the RT-LAMP amplified products with SYBR Green I dye can also be measured and quantified at the excitation wavelength of 498 nm and emission wavelength of 522 nm.

3.8.3 Real-Time Detection

1. Prepare the reaction mix using the components shown in Table 2, and follow Subheading 3.1 until **step 5**.
2. Place the reaction tubes in the hot plate of real-time turbidity meter LA-200, and incubate them at 63 °C for 90 min, followed by a 2 min enzyme heat inactivation step at 80 °C (*see* **Note 21**).

3.8.4 Interpretation of Positive and Negative Controls

1. Negative control tested positive: All samples tested are considered invalid since there is a high contamination probability. False positives cannot be ruled out.
2. Positive control tested negative: All samples tested are considered invalid since the RT-LAMP reaction may not have worked. False positives cannot be ruled out.

3.8.5 Disposal

1. RT-LAMP reactions must be disposed of in sealed bags. Never autoclave RT-LAMP reactions as this may aerosolize the RT-LAMP products and further result in false-positive reactions.

4 Notes

1. Synthesized LAMP primers are stored at −20 °C at 100 μM in stock aliquots.
2. A range of 80–120 mM $MgSO_4$ will work.

3. DNA polymerases that can be configured at room temperature are available. They offer the advantage of not needing ice or the equipment required to generate it. In contrast, most thermophilic DNA polymerases must be prepared on ice.
4. A reverse transcriptase enzyme is required for RNA templates to convert RNA to cDNA templates for LAMP amplification. There is a variety of commercial reverse transcriptase enzymes. If a different one than the one used in this protocol is used, it must be used in the concentration and working conditions established by the manufacturer.
5. Commercially available RNA extraction kits such as Qiagen RNeasy minikit and QiAamp DNA Blood mini-kit, or inhouse-based protocols can be used for HEV RT-LAMP [22, 24, 26, 28].
6. To reduce the risk of contamination, reagent preparation and experimentation should be performed in an environment free of amplified DNA and other nucleic acid templates. To reduce the probability of contamination, it is advisable to use a class 2 biological safety cabinet to prepare reagents, primers, and assays. Negative environmental controls should be used to monitor contamination. It is advisable to use only sterile plastic material and filter tips, as well as dividing the laboratory areas and the sets of pipettes used in LAMP assays.
7. It is advisable to make work aliquots of the primers and enzymes (at the concentration used in the reaction) to avoid repeating freezing and thawing cycles.
8. For a large number of samples, it will be enough to prepare a 5% excess master mix. If the analyses are carried out by an experienced technician, a 2% excess master mix will be sufficient.
9. The negative reaction control (NTC) includes all the PCR components except the cDNA template. It ensures the reliability of the experiment and the absence of contamination in the reagents or cross-contamination between samples. If negative reaction control (NTC) produces amplification, cDNA may be contaminated due to amplicon contamination in the RT reagents or contamination in the cDNA preparation. In this case, repeat the PCR reaction using new reagents and keep only the cDNA extraction. If this issue persists, it will be necessary to make a new RNA extraction.
10. When calculating to prepare the master mix, ensure that the number of samples includes a negative control, a positive control, a specificity control (optional), an environmental contamination control, and an additional amount to allow pipetting mistakes.

11. Enzymatic digestion is performed to confirm that the RT-LAMP products had the corresponding sequences of the selected target. The expected sizes of the fragments produced by BsiEI digestion are 208, 150, and 136 bp.
12. Heating must be monitored so that the liquid does not come out.
13. GreenSafe is an organic compound that forms part of cyanines. It can bind to nucleic acids and emit fluorescence. It has a maximum emission at 525 nm (corresponding to the green zone of the spectrum, hence its name). Any nucleic acid stain binds to nucleic acids and therefore is a carcinogen. Wear gloves when handling; appropriate safety measures should still be used.
14. Ensure no bubbles are trapped under the combs, and all bubbles on the gel surface are removed before the gel sets.
15. The electrophoresis buffer (TAE 1×) can be reused up to five assays.
16. The dyes in the loading buffer help monitor the migration of cDNA fragments depending on their size. The glycerol presence in the loading buffer solution increases the density of the sample, which is deposited at the bottom of the well. The loading buffer concentration is 6×. It is used at 1×, so for each 25 μL of amplified sample, add 5 μL of loading buffer.
17. When the sample is deposited at the bottom of the well, do not release the pipette until the tip has been removed from the well to avoid dispersion of the sample.
18. The voltage must be adjusted according to the size of the gel, especially according to its length and the electrophoresis cuvette used.
19. Observe the position where the DNA bands have migrated. To estimate the size of the DNA bands, compare their migration with that of the molecular weight standard (MW marker). The NZYDNA Ladder VI provides 17 bands ranging from 50 to 1500 bp. The speed of migration of DNA molecules in the gel is inversely proportional to their size.
20. The enzyme is visual endpoint detection inactivated by heating the tubes at 80–95 °C for 2–5 min. Turbidity detection is not affected by the presence of this reagent.
21. The real-time turbidity meter LA-200 measures each sample's turbidity every 6 s for 90 min. The sample is considered positive when a turbidity equal to or higher than 0.1 is achieved [29].

References

1. Notomi T, Mori Y, Tomita N, Kanda H (2015) Loop-mediated isothermal amplification (LAMP): principle, features, and future prospects. J Microbiol 53(1):1. https://doi.org/10.1007/S12275-015-4656-9
2. Salinas NR, Little DP (2012) Electric LAMP: virtual Loop-mediated isothermal AMPlification. ISRN Bioinform 2012:1–5. https://doi.org/10.5402/2012/696758
3. Santiago-Felipe S, Tortajada-Genaro LA, Carrascosa J, Puchades R, Maquieira Á (2016) Real-time Loop-mediated isothermal DNA amplification in compact disc micro-reactors. Biosens Bioelectron 79:300–306. https://doi.org/10.1016/J.BIOS.2015.12.045
4. Wong YP, Othman S, Lau YL, Radu S, Chee HY (2018) Loop-mediated isothermal amplification (LAMP): a versatile technique for detection of micro-organisms. J Appl Microbiol 124(3):626–643. https://doi.org/10.1111/JAM.13647
5. Xia Y, Guo XG, Zhou S (2014) Rapid detection of *Streptococcus pneumoniae* by real-time fluorescence Loop-mediated isothermal amplification. J Thorac Dis 6(9):1193. https://doi.org/10.3978/J.ISSN.2072-1439.2014.07.29
6. Nakano R et al (2015) Rapid detection of the *Klebsiella pneumoniae* carbapenemase (KPC) gene by Loop-mediated isothermal amplification (LAMP). J Infect Chemother 21(3):202–206. https://doi.org/10.1016/J.JIAC.2014.11.010
7. Fan F, Du P, Kan B, Yan M (2015) The development and evaluation of a Loop-mediated isothermal amplification method for the rapid detection of *Salmonella enterica* serovar Typhi. PLoS One 10(4). https://doi.org/10.1371/JOURNAL.PONE.0124507
8. Birmpa A, Kalogeropoulos K, Kokkinos P, Vantarakis A (2015) Evaluation of two Loop-mediated isothermal amplification methods for the detection of *Salmonella* Enteritidis and *Listeria monocytogenes* in artificially contaminated ready-to-eat fresh produce. Ital J Food Saf 4(3). https://doi.org/10.4081/IJFS.2015.5383
9. Zhang Y et al (2017) Development of Loop-mediated isothermal amplification with *Plasmodium falciparum* unique genes for molecular diagnosis of human malaria. Pathog Glob Health 111(5):247–255. https://doi.org/10.1080/20477724.2017.1347379
10. Britton S, Cheng Q, Grigg MJ, William T, Anstey NM, McCarthy JS (2016) A sensitive, colorimetric, high-throughput Loop-mediated isothermal amplification assay for the detection of *Plasmodium knowlesi*. Am J Trop Med Hyg 95(1):120. https://doi.org/10.4269/AJTMH.15-0670
11. Lau YL, Lai MY, Fong MY, Jelip J, Mahmud R (2016) Loop-mediated isothermal amplification assay for identification of five human plasmodium species in Malaysia. Am J Trop Med Hyg 94(2):336. https://doi.org/10.4269/AJTMH.15-0569
12. Ziros PG, Kokkinos PA, Allard A, Vantarakis A (2015) Development and evaluation of a Loop-mediated isothermal amplification assay for the detection of adenovirus 40 and 41. Food Environ Virol 7(3):276–285. https://doi.org/10.1007/S12560-015-9182-8
13. Niihara H, Kohno K, Tobita R, Ishikawa N, Morita E (2017) Rapid diagnosis of herpes zoster by Loop-mediated isothermal amplification in a pregnant woman showing folliculitis-like eruption without vesicles. J Dermatol 44(7):e174–e175. https://doi.org/10.1111/1346-8138.13861
14. Wang X et al (2015) Rapid detection of active human cytomegalovirus infection in pregnancy using Loop-mediated isothermal amplification. Mol Med Rep 12(2):2269–2274. https://doi.org/10.3892/MMR.2015.3572
15. Zhao N, Liu JX, Li D, Sun DX (2016) Loop-mediated isothermal amplification for visual detection of hepatitis B virus. Zhonghua Gan Zang Bing Za Zhi 24(6):406–411. https://doi.org/10.3760/CMA.J.ISSN.1007-3418.2016.06.003
16. Zhao N, Liu J, Sun D (2017) Detection of HCV genotypes 1b and 2a by a reverse transcription Loop-mediated isothermal amplification assay. J Med Virol 89(6):1048–1054. https://doi.org/10.1002/JMV.24747
17. Mori Y, Kanda H, Notomi T (2013) Loop-mediated isothermal amplification (LAMP): recent progress in research and development. J Infect Chemother 19(3):404–411. https://doi.org/10.1007/S10156-013-0590-0
18. Zhao HB et al (2014) Development of Loop-mediated isothermal amplification (LAMP) for universal detection of enteroviruses. Indian J Microbiol 54(1):80. https://doi.org/10.1007/S12088-013-0399-7
19. Lu L, Li C, Hagedorn CH (2006) Phylogenetic analysis of global hepatitis E virus sequences: genetic diversity, subtypes and

zoonosis. Rev Med Virol 16(1):5–36. https://doi.org/10.1002/RMV.482

20. Kamar N et al (2012) Hepatitis E. Lancet 379(9835):2477–2488. https://doi.org/10.1016/S0140-6736(11)61849-7
21. Aggarwal R, Naik S (2009) Epidemiology of hepatitis E: current status. J Gastroenterol Hepatol 24(9):1484–1493. https://doi.org/10.1111/J.1440-1746.2009.05933.X
22. Zhang LQ et al (2012) Simple and rapid detection of swine hepatitis E virus by reverse transcription Loop-mediated isothermal amplification. Arch Virol 157(12):2383–2388. https://doi.org/10.1007/S00705-012-1425-5
23. Chen Q, Yuan L, Wan J, Chen Y, Du C (2014) Colorimetric detection of hepatitis E virus based on reverse transcription loop mediated isothermal amplification (RT-LAMP) assay. J Virol Methods 197:29–33. https://doi.org/10.1016/J.JVIROMET.2013.11.006
24. Gao S et al (2016) Development and evaluation of a RT-LAMP assay for rapid detection of hepatitis E virus from shellfish. Int J Food Microbiol 220:1–5. https://doi.org/10.1016/J.IJFOODMICRO.2015.12.008
25. Wang J, Zhang H, Yan S, Wang Y, Lu X, Fu H (2017) Development of a one step reverse transcription-Loop-mediated isothermal amplification system for a highly sensitive detection of rabbit hepatitis E virus. Clin Lab 63(5):901–905. https://doi.org/10.7754/CLIN.LAB.2017.160919
26. Lan X, Yang B, Bao YL, Xiang PY, Xue RL, Ji XL (2009) Reverse transcription-Loop-mediated isothermal amplification assay for rapid detection of hepatitis E virus. J Clin Microbiol 47(7):2304. https://doi.org/10.1128/JCM.00498-09
27. Saxena A et al (2022) Development and clinical validation of RT-LAMP-based lateral-flow devices and electrochemical sensor for detecting multigene targets in SARS-CoV-2. Int J Mol Sci 23(21). https://doi.org/10.3390/IJMS232113105
28. Nyan DC, Swinson KL (2015) A novel multiplex isothermal amplification method for rapid detection and identification of viruses. Sci Rep 5. https://doi.org/10.1038/SREP17925
29. Boubourakas IN, Kyriakopoulou PE (2022) Reverse transcription Loop-mediated isothermal amplification (RT-LAMP) at the service of viroid detection. Methods Mol Biol 2316:195–210. https://doi.org/10.1007/978-1-0716-1464-8_17

Chapter 13

Detection of Adenovirus by LAMP

Kotsiri Zoi and Vantarakis Apostolos

Abstract

Adenovirus is transmitted through food with lower impact in humans. It is considered as a cause of foodborne disease in humans. Following an 8–10-day incubation period, symptoms such as diarrhea, vomiting, dehydration, and fever appear and persist for 7–8 days. The virus does not start to shed until 7–14 days after infection. There are no case reports of viruses being clearly transmitted through food. Adenoviruses can be transmitted by direct contact between individuals, by sneezing and coughing, by touching contaminated surfaces, or through the fecal-oral route. Transmission of adenoviruses can happen in water. Analytical techniques that can shorten turnaround times while preserving sensitivity and specificity can be used to control risk. Here, we discuss innovative methods for detecting adenoviruses, like the LAMP method, that shorten the time required for a response and may lower the likelihood of foodborne outbreaks.

Key words Foodborne pathogen, Food analysis, Molecular method, Adenovirus, Loop-mediated isothermal amplification (LAMP), Food

1 Introduction

1.1 Adenovirus

Adenoviruses are double-stranded DNA viruses found in a plethora of organisms and are considered as causative agent in mammals, birds, and amphibia [1]. Adenoviruses are nonenveloped and consist of 26–45 kb double-stranded DNA viruses with diameters of 90–100 nm. Their host is an icosahedral capsid, which is made up primarily of three parts: fiber, hexon, and penton base [2, 3].

The Adenoviridae family comprises five genera: *Atadenovirus* (8 species), which infects birds, lizards, and mammals; *Aviadenovirus* (14 species), which infects birds; *Ichtadenovirus* (1 species), isolated from sturgeon; *Mastadenovirus* (45 species), which infects mammals only, including humans; and *Siadenovirus* (6 species), which primarily infects birds and one species of frog [4].

The transmission route is mainly via contact with contaminated surfaces; coughing, sneezing, or the fecal-oral route are the ways that adenoviruses are transmitted between individuals [3,4]. Studies

Alejandro Garrido-Maestu and Alexandre Lamas (eds.), *Foodborne Pathogens: Loop-mediated Isothermal Amplification Protocols*, Methods and Protocols in Food Science, https://doi.org/10.1007/978-1-0716-4526-0_13,

have shown that adenovirus-associated gastroenteritis commonly affects populations in settings like medical facilities, military bases, and educational institutions [4].

Human adenovirus (HAdV) is frequently found in drinking water and is primarily linked to gastroenteritis contracted by water [5]. Adenovirus is recorded, between other enteric pathogens, most commonly found in contaminated freshwaters [6–8]. It is considered as one of the nine microorganisms on the drinking water contamination candidate list compiled by the United States Environmental Protection Agency (USEPA) [9].

Although reports of adenovirus-related drinking water-borne disease outbreaks are still pending, the discovery of these viruses in treated drinking water sources raises the possibility of exposure [4]. There are currently no case reports or epidemiological data demonstrating the viruses' clear foodborne transmission. Therefore, the possibility of human respiratory viruses spreading through foodborne means is still present [2].

Few cases of adenovirus infection in fish have been documented, though it does occur occasionally. The Atlantic cod and dab exhibit epidermal hyperplasia, while Japanese red sea bream also exhibit hyperplasia. *Acipenser transmontanus*, the white sturgeon, is the only known adenovirus disease. White sturgeon adenovirus 1 (WSAdV-1) is linked to this chronic infection [10].

Moreover, symptoms like diarrhea, vomiting, dehydration, and fever have been documented after an 8–10-day incubation period. The virus does not start to shed until 7–14 days after infection. Median tissue culture infective dose was calculated as 3.2 $TCID_{50}$/mL [11].

According to the ECDC report of 2021, the most recent report, there was only one case of contamination with adenovirus as causative agent in Poland [12]. Failures in chlorination are frequently mentioned as a primary cause of outbreaks [13].

1.2 Diagnostic Methods

Many diagnostic methods have been described, such as antibody detection, antigen detection techniques, culture, electron microscopy, PCR, in situ hybridization, immunohistochemistry, and histopathology. A few of these are not frequently utilized. For example, serology is not very useful because of the low humoral immune response in many high-risk patients as well as the high population seroprevalence of adenovirus in the absence of active infection. Nucleic acid sequence-based techniques or immunologic approaches can be used to carry out viral subtyping and serotyping. Currently, epidemiological studies and other research applications are the main uses for these techniques. Loop-mediated isothermal amplification (LAMP) is considered as a sensitive, robust, and low-cost technique. In terms of comparability, LAMP is a simpler approach to viral detection than PCR because it allows for visual confirmation of the results and simple temperature control during the reaction [14].

1.3 Applications

The identification and measurement of adenovirus in food have not been thoroughly documented. These mostly emphasize on alimentary raw foods including lettuce, strawberries, and cherry tomatoes and shellfish [3, 15, 16]. In order to detect adenoviruses, research focused on virus concentration in a limited volume use PCR. There hasn't been any research done on the LAMP method's viral detection on food products. With the onset of the pandemic and the advancement of technology, there is still more space for the virus to be updated via quick and precise techniques [17].

2 Materials

Use only pure water and reagents of the highest caliber for preparing all solutions. Assemble and maintain all reagents at room temperature. When disposing of waste materials, take care to adhere to all waste disposal regulations.

2.1 Tris/Glycine/Beef Extract (TGBE) Buffer

1. Weigh 12.1 g Tris base, 3.8 g glycine, 10 g beef extract powder in to a volumetric flask. Add water to a volume of 1000 mL. Mix and adjust pH to 9.5 at 25 °C.

2.2 5 × PEG/NaCl Solution (500 g/L PEG 8000, 1.5 Mol/L NaCl)

1. Weigh 500 g polyethylene glycol (PEG) 8000, 87 g sodium chloride (NaCl), into a volumetric flask. Add water to a volume of 1000 mL. Mix and adjust pH to 9.5 at 25 °C.
2. Dissolve the solids in 450 mL water, heating gently if necessary.
3. Adjust volume to 1000 mL with water and mix well. Sterilize by autoclaving.

2.3 Phosphate-Buffered Saline (PBS)

1. Weigh 8 g sodium chloride, 0.2 g potassium chloride, 1.15 g disodium hydrogen phosphate, and 0.2 g potassium dihydrogen phosphate into a volumetric flask. Add water to a volume of 1000 mL. Mix and adjust pH to 7.3 at 25 °C.

2.4 Chloroform/Butanol Mixture (1:1 Volume Fraction)

1. Weigh 10 mL chloroform and 10 mL n-butanol into a volumetric flask. Mix them gently.

2.5 Tryptose Phosphate Broth

According to manufacturer instructions.

2.6 Polyethylene Glycol (PEG) 8000

1. Weigh 1.2 gr polyethylene glycol into a volumetric flask. Mix them gently.

2.7 NucliSens® Lysis Buffer and NucliSens® (Biomerieux)

According to manufacturer instructions.

2.8 Primers for the Detection of Adenovirus

1. In the following table, the sequence of primers is presented.

Table 1
Primers used in for adenovirus determination in LAMP tests

Primers	Sequence
F3_ADE	CCGCAATGGTCTTACATGCA
B3_ADE	GAATAAGCGGTGTCCTCGC
FIP_ADE	CTGAAGTACGTATCGGTGGCGCGACGCCTCGGAGTA TCTGAG
BIP_ADE	TCAGAAATCCCACTGTGGCTCCCGAATCGCAGCG TCAGTC
LF_ADE	AATTGCACCAGGCCCGG
LB_ADE	CACGATGTAACCACAGACAGGTC

2.9 Bst2.0 DNA Polymerase Reaction Buffer (New England Biolabs, MA)

2.10 Bst2.0 DNA Polymerase (8 U) (New England Biolabs, MA)

2.11 $MgSO_4$ (100 mM)

2.12 Betaine (5 M) (Sigma Aldrich, St. Louis)

2.13 dNTPs (1.2 mM)

2.14 SYBR Green

2.15 WarmStart Colorimetric LAMP 2× Master Mix (New England Biolabs)

2.16 Nuclease-Free Water

3 Methods

3.1 Condensation Step

3.1.1 Soft Fruits and Vegetables [16]

1. Cut 25 g of soft fruits or vegetables into pieces and transfer to the sample in a 400 mL mesh filter bag.
2. Add 40 mL TGBE.
3. Incubate at room temperature with constant rocking at approximately 60 oscillations/min for 20 min.
4. Adjust pH to 9.5 ± 0.5 with NaOH (≥ 10 mol/L). Do not make more than three such pH adjustments per sample.
5. Decant the eluate from the filtered compartment into a centrifuge tube.
6. Clarify by centrifugation at 10.000× *g* for 30 min at 5 °C.
7. Decant the supernatant into a single clean tube or bottle and adjust to pH 7.0 ± 0.5 with HCl (≥ 5 mol/L).
8. Add 0.25 volumes of 5 × PEG/NaCl solution, and produce a final concentration of 100 g/L PEG 0.3 mol/L NaCl.
9. Homogenize by shaking for 60 s and then incubate with constant rocking at approximately 60 oscillations/min at 5 °C for 60 min.
10. Centrifuge at 10.000× *g* for 30 min at 5 °C.
11. Decant and discard the supernatant, and then centrifuge at 10.000× *g* for 5 min at 5 °C to compact the pellet.
12. Discard the supernatant and resuspend the pellet in 500 μL PBS.
13. Transfer the suspension to a chloroform-resistant centrifuge tube.
14. Add 500 uL chloroform/butanol mixture, vortex to mix, and then incubate at room temperature for 5 min.

15. Centrifuge at 10.000× *g* for 15 min at 5 °C.
16. Carefully transfer the aqueous phase to a fresh tube and retain for DNA extraction.

3.1.2 Shellfish [18]

1. Remove with a sterile oyster knife for digestive tissues the hepatopancreas.
2. Collect 2 g of hepatopancreas and homogenate them (tissues derived from 12 dissected oysters).
3. Resuspend the mix in 10 mL of tryptose phosphate broth dissolved in 0.05 mol/L glycine pH 9.
4. The homogenate was purified with an equal volume of chloroform–butanol (1:1).
5. Centrifuge at 13500× *g* at 4 °C for 15 min.
6. Precipitate the supernatant with 12% polyethylene glycol.
7. Dissolve the liquid in 0.75 mol/L NaCl for 2 h at 4 °C.
8. Centrifuge at 13500× *g* and 4 °C for 20 min. Each pellet was resuspended in 3 mL of ultrapure water and clarified twice with 30% of chloroform.

3.2 DNA Extraction

For extraction from soft fruit, a further extraction step is required. According to manufacturer instructions, NucliSens® lysis buffer and NucliSens® magnetic extraction reagents (bioMerieux) were used to extract nucleic acids. In this kit, magnetic silica solution; wash buffers 1, 2, and 3; and elution buffer are included. The final elution using 150 μL elution buffer was carried out with samples from the soft fruit and salad vegetable supply chains, yielding a 300 μL nucleic acid extract. The nucleic acid extract was either tested right away or kept in storage at −70 °C (*see* **Note 1**).

3.3 Amplification

The amplification with SYBR green and colorimetric LAMP was used the same primers.

3.3.1 LAMP SYBR Green

LAMP primers (Table 1) are included in 20 μL reaction mixture as follows:

1. Mix the reagents gently referred in Table 2. (*see* **Notes 2** and **3**).
2. Add 1000× SYBR green on inner top of tube.
3. Incubate them at 69 °C for 60 min.
4. Centrifuge tube to mix the solution with SYBR green.
5. Visualize the results under a UV lamp.

Table 2
LAMP reagents with their corresponding concentrations

Reagents	Stock	Final concentration
LAMP buffer mix	10× ($MgSO_4$ 2 mM)	1×
$MgSO_4$	100 mM	7 mM
Primer mix F3/B3 FIP/BIP LF/LB	10× 2 mM 16 mM 8 mM	1× 0.2 mM 1.6 mM 0.8 mM
dNTPs	10 mM	1.2 mM
Betaine	5 M	0.6 M
BS2.0 enzyme	8 units/μL	8 units
Sample		–
H_20		–
SYBR green	1000x	100 mM

Table 3
Colorimetric LAMP reagents with their corresponding concentrations

Reagents	Stock	Concentration
LAMP colorimetric	2×	1×
Primer mix 10× LB- LF B3-F3 BIP-FIP	10× 4 mM 16 mM 2 mM	1× 0.4 mM 1.6 mM 0.2 mM
Sample	–	–
H_20		

3.3.2 LAMP Colorimetric

LAMP primers (Table 1) are included in 20 μL reaction mixture as follows:

1. Mix the reagents gently referred in Table 3 (*see* **Notes 2** and **3**).
2. Incubate them at 64 °C for 60 min.
3. Visualize the results by color change (pink to yellow) (Fig. 1).

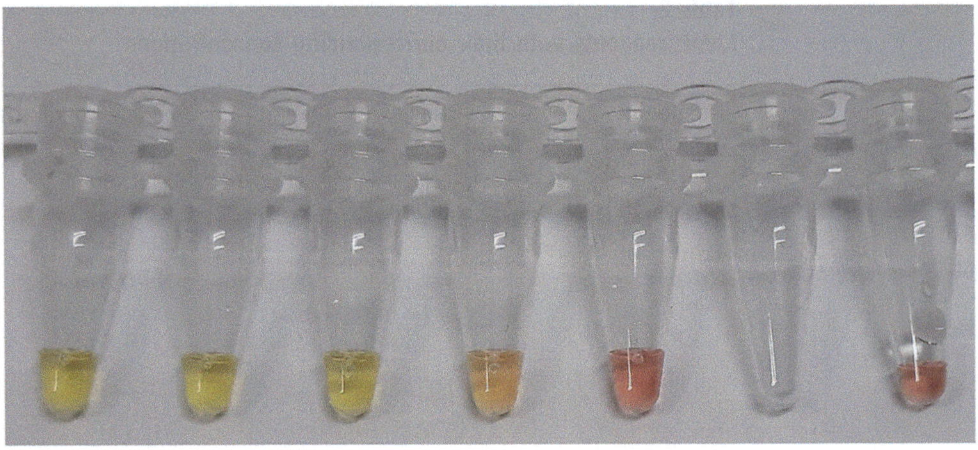

Fig. 1 LAMP colorimetric dilutions (from left to right yellow to orange as positive samples and pink as negatives)

4 Notes

1. Extracted DNA must be processed right away or kept for longer periods of time at −70 °C or below, at −15 °C or below for up to 6 months, or at 5 ± 3 °C for a maximum of 24 h.
2. A negative DNA extraction control must be included for every batch of samples that are tested.
3. A positive DNA extraction control must be included for every batch of samples that are tested.

References

1. World Health Organization (2011) Guidelines for drinking-water quality, 4th edn. WHO Press, Geneva
2. Brien BO, Goodridge L, Ronholm J, Nasheri N (2020) Since January 2020 Elsevier has created a COVID-19 resource centre with free information in English and Mandarin on the novel coronavirus COVID- 19. The COVID-19 resource centre is hosted on Elsevier Connect, the company ' s public news and information
3. D'Souza DH (2015) Update on foodborne viruses: types, concentration and sampling methods. Woodhead Publishing Limited
4. O'Shea H, Blacklaws BA, Collins PJ et al (2019) Viruses associated with foodborne infections. Ref Modul Life Sci. https://doi.org/10.1016/b978-0-12-809633-8.90273-5
5. Rashid M, Khan MN, Jalbani N (2021) Detection of human adenovirus, rotavirus, and enterovirus in tap water and their association with the overall quality of water in Karachi, Pakistan. Food Environ Virol 13:44–52. https://doi.org/10.1007/s12560-020-09448-8
6. Weaver L, Abraham P, Pang L et al (2023) Comparative reductions of norovirus, echovirus, adenovirus, campylobacter jejuni and process indicator organisms during water filtration in alluvial sand. Sci Total Environ 888:164178. https://doi.org/10.1016/j.scitotenv.2023.164178
7. Pang X, Qiu Y, Gao T et al (2019) Prevalence, levels and seasonal variations of human enteric viruses in six major rivers in Alberta, Canada. Water Res 153:349–356. https://doi.org/10.1016/j.watres.2019.01.034

8. Haramoto E, Kitajima M, Hata A et al (2018) A review on recent progress in the detection methods and prevalence of human enteric viruses in water. Water Res 135:168–186. https://doi.org/10.1016/j.watres.2018.02.004
9. Jiang SC (2006) Human adenoviruses in water: occurrence and health implications: a critical review. Environ Sci Technol 40:7132–7140. https://doi.org/10.1021/es060892o
10. Nagy E (2016) Adenoviruses of fish. Elsevier Inc.
11. Mikszewski A, Stabile L, Buonanno G, Morawska L (2022) The airborne contagiousness of respiratory viruses: a comparative analysis and implications for mitigation. Geosci Front 13:101285. https://doi.org/10.1016/j.gsf.2021.101285
12. Food E, Authority S (2022) The European Union one health 2021 Zoonoses report. EFSA J 20. https://doi.org/10.2903/j.efsa.2022.7666
13. Rodríguez-Lázaro D, Cook N, Ruggeri FM et al (2012) Virus hazards from food, water and other contaminated environments. FEMS Microbiol Rev 36:786–814. https://doi.org/10.1111/j.1574-6976.2011.00306.x
14. Kimura Y, Ikeuchi M (2023) Development of non-electrically controlled SalivaDirect LAMP (NEC-SD-LAMP), a new nonelectrical infectious disease testing method. Sci Rep 13:1–11. https://doi.org/10.1038/s41598-023-38800-8
15. Carratalà A, Rodriguez-Manzano J, Hundesa A et al (2013) Effect of temperature and sunlight on the stability of human adenoviruses and MS2 as fecal contaminants on fresh produce surfaces. Int J Food Microbiol 164:128–134. https://doi.org/10.1016/j.ijfoodmicro.2013.04.007
16. Birmpa A, Bellou M, Kokkinos P, Vantarakis A (2016) Effect of non-thermal, conventional, and combined disinfection technologies on the stability of human adenoviruses as fecal contaminants on surfaces of fresh ready-to-eat products. J Food Prot 79:454–462. https://doi.org/10.4315/0362-028X.JFP-15-013
17. Roberts A, Chouhan RS, Shahdeo D et al (2021) A recent update on advanced molecular diagnostic techniques for COVID-19 pandemic: an overview. Front Immunol 12:1–16. https://doi.org/10.3389/fimmu.2021.732756
18. Rigotto C, Victoria M, Moresco V et al (2010) Assessment of adenovirus, hepatitis a virus and rotavirus presence in environmental samples in Florianopolis, South Brazil. J Appl Microbiol 109:1979–1987. https://doi.org/10.1111/j.1365-2672.2010.04827.x

Part IV

Other Microbes

Chapter 14

Detection of Spoilage Fungi in Foods by Loop-Mediated Isothermal Amplification

Foteini Roumani, Saioa Gómez, Marta Prado, and Alejandro Garrido-Maestu

Abstract

Molecular DNA-based methods, such as PCR/qPCR, have been employed to detect yeasts and molds in various food products, offering faster results due to their heightened sensitivity and specificity, compared to culture-based techniques. However, recently, isothermal amplification techniques like loop-mediated isothermal amplification (LAMP) have emerged, simplifying assays and reducing costs. In this chapter a panfungal LAMP assay (LAMP-18S) for the rapid detection of spoilage responsible fungi in fruit preparations is described. Primers targeting the 18S rRNA gene were designed for this purpose. The study explored two distinct detection strategies: real-time fluorescence and naked-eye color detection.

Key words LAMP, Fungi, Colorimetric detection, Naked eye, Real-time fluorescence

1 Introduction

Traditional methods for detecting spoilage responsible fungi rely on culture-based approaches, specifically pour plating in differential and selective media. However, these methods are known to be time-consuming and labor-intensive [1, 2]. The extended duration, often up to 7 days, makes them impractical for products with short shelf-lives. Additionally, these techniques are retrospective in nature, offering limited value for real-time quality and process control [2]. Nevertheless, lately, DNA amplification techniques have emerged as an alternative to address the limitations of traditional methods, owing to their high sensitivity and specificity. Polymerase chain reaction (PCR) and real-time PCR (qPCR) have become prominent in the detection of spoilage yeasts and molds across various food products [3–7].

Recently, in contrast to PCR/qPCR, isothermal nucleic acid amplification techniques have emerged as an alternative, offering simplified assays and reduced costs. These techniques operate at a

Alejandro Garrido-Maestu and Alexandre Lamas (eds.), *Foodborne Pathogens: Loop-mediated Isothermal Amplification Protocols*, Methods and Protocols in Food Science, https://doi.org/10.1007/978-1-0716-4526-0_14,

constant temperature, eliminating the need for expensive equipment, as a simple heat block or water bath is sufficient. Among various isothermal amplification methods, the loop-mediated isothermal amplification (LAMP), developed by Notomi et al. [8], has gained significant popularity due to its high specificity, attributed to the requirement for 4–6 primers in the assay. Additionally, the results can be interpreted in various ways, including naked-eye observation of turbidity or color change [9, 10]. Moreover, LAMP exhibits greater resistance to many common PCR inhibitors, such as polyphenols and polysaccharides, which can be abundant in diverse food matrices [11]. This resilience enhances the applicability of LAMP in challenging sample environments.

While LAMP has found extensive application in detecting pathogenic bacteria [12–14] and viruses [15–17], its use for the detection of foodborne fungi is relatively constrained. Primarily, its application in this context is centered around the detection of mycotoxigenic fungi [9, 18, 19]. Additionally, two panfungal assays have been reported one for detecting contamination in peppers and paprika powder [20] and the other for environmental sampling [21]. However, to our knowledge, it has not been developed a panfungal LAMP assay coupled with colorimetric naked-eye detection for food samples. In this sense, in the present chapter, a new panfungal LAMP assay is described. For this reason, the selected target gene was the 18S rRNA gene. Two different detection chemistries were chosen, namely, real-time fluorescence and naked-eye color change. The described methodology can be applied for the detection of fungi in various food products.

2 Materials

2.1 Media

1. Malt Extract Broth (MEB, Liofilchem S.r.l., Italy): 6.0 g malt extract, 1.8 g maltose, 6.0 g glucose, 1.2 g yeast extract, 1000 mL Milli-Q water, pH 4.7 ± 0.2.
2. Potato Dextrose agar (PDA, Biokar diagnostics S.A., France): 4.0 g potato extract, 20.0 g glucose, 15.0 g bacteriological agar, 1000 mL Milli-Q water, pH 5.6 ± 0.2.
3. Dichloran Rose Bengal Chloramphenicol agar (DRBC, Biokar diagnostics S.A., France): 5.0 g polypeptone, 10.0 g glucose, 1.0 g monopotassium phosphate, 0.5 g $MgSO_4.H_2O$, 2.0 mg dichloran, 25.0 mg rose Bengal, 50.0 mg chloramphenicol, 50.0 mg chlortetracycline chlorhydrate, 10.0 mg $ZnSO_4.7H_2O$, 5.0 mg $CuSO_4.5H_2O$, 1 mL tergitol, 12.4 g bacteriological agar, 1000 mL Milli-Q water, pH 5.6 ± 0.2.

Table 1
LAMP primers list

Prime/ probe	Assay	Sequence (5′ → 3′)	Modification	References
F3 18S	LAMP	ACG GGG AAT AAG GGT TCG A	–	[23]
B3 18S	18S	TTG GAG CTG GAA TTA CCG C	–	
FIP 18S		ATT GGG TAA TTT GCG CGC CTG TTT TAG AGG GAG CCT GAG AAA CG	–	
BIP 18S		TAC AGG GCC CTT TCG GGT CTT TTT CTT GCC CTC CAA TTG TTC CT	–	
LF 18S		GCC TTC CTT GGA TGT GGT AGC	–	

Primers were designed using the Primer Explorer V4 (http://primerexplorer.jp/elamp4.0.0/index.html) and purchased from Integrated DNA Technologies Inc. (IDT, Leuven, Belgium)

2.2 Chemicals

1. Phosphate Buffer Saline (1× PBS): 8 g NaCl, 0.2 g KCl, 1.44 g Na_2HPO_4, 0.245 g KH_2PO_4, 1000 mL Milli-Q water, pH 7.4 (*see* **Note 1**).
2. DNeasy PowerSoil Pro kit (Qiagen, Barcelona, Spain).
3. Qubit™ 1× dsDNA HS Assay Kit (Invitrogen™, Thermo-Fisher Scientific, Carlsbad, CA, USA).
4. LAMP primers (see Table 1).
5. GspSSD Isothermal Mastermix (Optigene Ltd., Horsham, UK) for fluorescent assays.
6. Visual detection RT Isothermal Mastermix (Optigene Ltd., Horsham, UK) for colorimetric assays (*see* **Note 2**).

2.3 Equipment

1. Stomacher or any other homogenization device. In the present assay, Stomacher® 400 Circulator (Seward Limited, West Sussex, UK) was used.
2. Shake incubator with constant agitation for sample enrichment.
3. Centrifuge with a speed range of 300–16,000× *g*, equipped with rotor suitable for 1.5–2 mL tubes.
4. Vortex with adaptor for tubes.
5. Qubit™ 4 fluorometer (Invitrogen™, Carlsbad, CA, USA). Suitable 0.5 mL tubes must also be purchased.
6. Real-time thermocycler. In the present assay, StepOne Plus™-Real-Time PCR system (Applied Biosystems™, Foster City, CA, USA) with StepOne™Software v2.3 was used. Experiments can be conducted in 0.1 mL PCR tubes or strips, as well as in 96-well plates, sealed using either optical-quality adhesive film or plastic lids.

7. Conventional thermocycler. In the described assay, Veriti Thermal Cycler (Applied Biosystems™, Foster City, CA, USA) was selected. The reactions run in 0.1 mL PCR tubes/strips with plastic lids (*see* **Note 3**).

3 Methods

3.1 Enrichment

3.1.1 Yeast Cultures

1. Inoculate a loopful from stock cultures in 5 mL tubes of MEB followed by incubation at 30 °C for 24 h.

3.1.2 Spore Cultures

1. Streak on PDA plates from stock cultures (−80 °C, 30% glycerol) and incubate for 7 days at 25 °C (*see* **Note 4**).
2. After the specified period, harvest the spores by scraping the surface of the plate with 10 mL of Milli-Q water.
3. Pass through a syringe filled with cotton to remove the hyphae.

3.1.3 Food Samples

1. Prepare the samples by weighing 50 g of sample in a stomacher bag with filter and mix with 50 mL of MEB.
2. Spike with the desire concentration of microorganisms and homogenize for 30 s in a Stomacher (*see* **Note 5**).
3. Incubate at 30 °C for 24 h with constant agitation at 150 rpm.
4. After the enrichment, perform confirmation plating by streaking on DRBC, followed by incubation at 30 °C for 24–96 h.

3.2 Sample Pretreatment

3.2.1 Fungal Cultures

1. Transfer 1 mL of fresh yeast or spore culture to a 2 mL Eppendorf tube.
2. Centrifuge at 9000× *g* for 5 min. Remove the supernatant.
3. Continue with the DNA extraction.

3.2.2 Food Samples

1. Recover 10 mL of sample from the filter side of the stomacher bag and transfer them in a clean 15 mL falcon tube.
2. Centrifuge at 440× *g* for 2 min.
3. Transfer the supernatant to a clean 15 mL falcon tube and centrifuge at 9000× *g* for 5 min. Discard the supernatant.
4. Resuspend the pellet with 1 mL of PBS and centrifuge again at 9000× *g* for 5 min. Discard the supernatant.
5. Continue with the DNA extraction.

3.3 DNA Extraction

The DNA extraction was carried out using the DNeasy PowerSoil Pro kit (Qiagen, Barcelona, Spain), following the standard protocol without any modifications.

1. Resuspend the pellet from the sample pretreatment (*see* Subheading 3.2) with 800 μL of Solution CD1. Transfer the suspension to a PowerBead Pro tube and vortex briefly.
2. Vortex the PowerBead Pro tube at maximum speed for 10 min, horizontally using a Vortex Adapter for 1.5–2 mL tubes.
3. Centrifuge the PowerBead Pro tube at 15.000× *g* for 1 min.
4. Transfer the supernatant to a clean 2 mL microcentrifuge tube.
5. Add 200 mL of Solution CD2 and vortex for 5 s.
6. Centrifuge at 15,000× *g* for 1 min. Avoiding the pellet, transfer up to 700 μL of supernatant to a clean 2 mL microcentrifuge tube.
7. Add 600 μL of Solution CD3 and vortex for 5 s.
8. Load 650 μL of lysate to an MB Spin Column. Centrifuge at 15,000× *g* for 1 min.
9. Discard the flow-through and repeat **step 8** to ensure that all the lysate has passed through the MB Spin Column.
10. Carefully place the MB Spin Column into a clean 2 mL Collection Tube. Avoid splashing any flow-through onto the MB Spin Column.
11. Add 500 μL of Solution EA to the MB Spin Column. Centrifuge at 15,000× *g* for 1 min.
12. Discard the flow-through and place the MB Spin Column back into the same 2 mL Collection Tube.
13. Add 500 μL of Solution C5 to the MB Spin Column. Centrifuge at 15,000× *g* for 1 min.
14. Discard the flow-through and place the MB Spin Column into a new 2 mL Collection Tube.
15. Centrifuge at up to 16,000× *g* for 2 min. Carefully place the MB Spin Column into a new 1.5 mL Elution Tube.
16. Add 50–100 μL of Solution C6 to the center of the white filter membrane.
17. Centrifuge at 15,000× *g* for 1 min. Discard the MB Spin Column. The DNA is now ready for downstream applications.
18. Store the DNA at −20 °C until further use.

3.4 DNA Quantification

The standard protocol for Qubit™ 1× dsDNA HS Assay Kit (Invitrogen™, Thermo-Fisher Scientific, Carlsbad, CA, USA) was followed.

1. Remove Qubit™ reagents from the refrigerator and allow them to reach room temperature. It is advisable to conduct the assay at room temperature (18–28 °C) to maintain accuracy, as temperature fluctuations may affect the results.

2. Prepare the standards and sample reactions. For standards add 190 μL of Qubit™ 1× dsDNA HS working solution to two Qubit™ 0.5 mL compatible tubes. Add 10 μL of Standard #1 (0 ng/μL) in one tube and 10 μL of Standard #2 (10 ng/μL) in another one. For the samples, 1–20 μL may be measured, and the working solution will be added to make a final volume of 200 μL. The amount of sample used depends on the sample volume available and the DNA concentration of the sample. In this protocol, 1 μL is used. Tubes should be labeled only in the cap and not in the walls as it can interfere with the reading.
3. Mix the standards and samples using a vortex for 5 s and incubate in the dark for at least 2 min. Reactions remain stable in the dark for up to 3 h.
4. Read the standards in the equipment to perform the calibration, and then proceed with the samples. Select in the equipment the volume of sample used in the reaction, as it is necessary to calculate the DNA concentration in the original sample. In case the sample result is "concentration too high," perform a 1/10 dilution of the original sample, and repeat the process starting from **step 2**. In case the equipment indicates "concentration too low," repeat the process, but this time increase the volume of the sample in **step 2**. If the problem persists, perform again the DNA isolation.

3.5 Real-Time Fluorescence LAMP Assay

1. The assay described below targets the 18S rRNA gene that has been used for universal detection of fungi. The sequences and concentrations of the primers are shown in Tables 1 and 2, respectively (*see* **Note 6**).

Table 2
Reagent volumes and concentration for one LAMP assay

Reagents	Volume/ rxn (μL)	Final concentration (nM)
Master Mix[a]	15	–
F3/B3 18S[b]	0.5	200
FIP/BIP 18S[c]	0.3	1000
LF 18S[b]	0.75	300
Template DNA	5	–
H_2O	2.8	–
Final volume	25	–

[a]GspSSD Isothermal Mastermix (fluorescent) or Visual detection RT Isothermal Mastermix (colorimetric)
[b]A stock concentration of 10,000 nM was used
[c]A stock concentration of 100,000 nM was used

2. Prepare tenfold dilutions of the DNA extract in Milli-Q water (e.g., 5 μL of DNA and 45 μL of H_2O).
3. Prepare the reaction mixture taking into consideration the number of reactions, the technical replicates (duplicates or ideally triplicates), as well as one negative control (Milli-Q water). Always consider one extra reaction to have enough volume. In Table 2 the volumes to be added per reaction are described.
4. The thermal profile of the reaction is described at Table 3 and for running the experiment a real-time thermocycler is needed.
5. A sample will be considered positive for fungal contamination if amplification is observed. In case no amplification is detected, this means that the sample is negative for fungi. In Fig. 1, typical amplification plots are shown.
6. By testing tenfold serial dilutions of pure fungal DNA, the dynamic range of the assay can be determined as the lowest concentration that can be detected by the method. Dynamic range can serve as an indicator of the analytical sensitivity of the assay.

Table 3
Thermal profile for fluorescent and colorimetric LAMP assay

Assay	Step[a]	Temperature (°C)	Time	Cycles
Fluorescent	Amplification	66	30 s	60[b]
	Melt-curve analysis	95	15 s	–
		80 up to 95 (0.2 °C increments)	60 s	
Colorimetric	Amplification	66	30 min[c]	–

[a]The step temperature and time may be modified depending on the selected Master Mix
[b]Fluorescence acquisition is performed at the end of each cycle
[c]Amplification time may be extended up to 40 min to observe better the color change

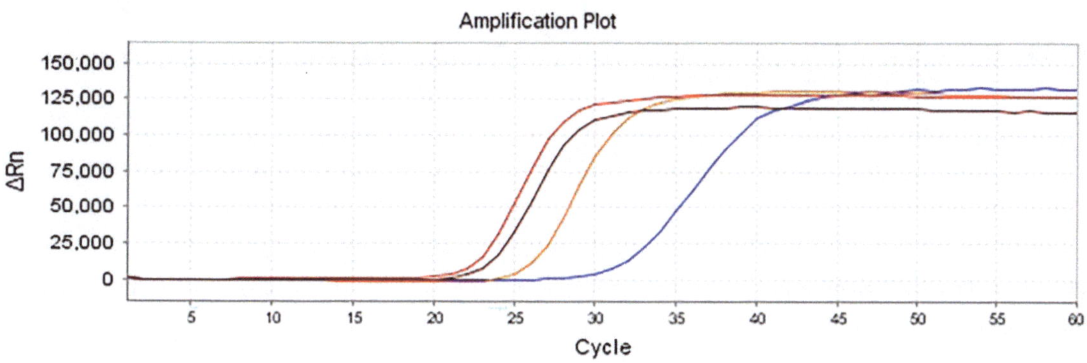

Fig. 1 Typical amplification plots of positive samples for the 18S rRNA gene

3.6 Colorimetric LAMP Assay

1. The assay targets the same gene as with the fluorescent LAMP and the same set of primers have been used as described above (*see* Tables 1 and 2).
2. Prepare tenfold dilutions of the DNA extract in Milli-Q water (e.g., 5 μL of DNA and 45 μL of H_2O).
3. Prepare the reaction mixture taking into consideration the number of reactions, the technical replicates (duplicates or ideally triplicates), as well as one negative control (Milli-Q water). Always consider one extra reaction to have enough volume. In Table 2 the volumes to be added per reaction are described.
4. For the assay a thermal block or any other heating device is required. The thermal profile of the assay is described in Table 3.
5. A sample turning light blue/turquoise is considered positive for fungal contamination, while a colorless/transparent sample is considered negative. In Fig. 2, examples of positive and negative samples are depicted.
6. By testing tenfold serial dilutions of pure fungal DNA, the dynamic range of the assay can be determined as the lowest concentration that can be detected by the method. Dynamic range can serve as an indicator of the analytical sensitivity of the assay.

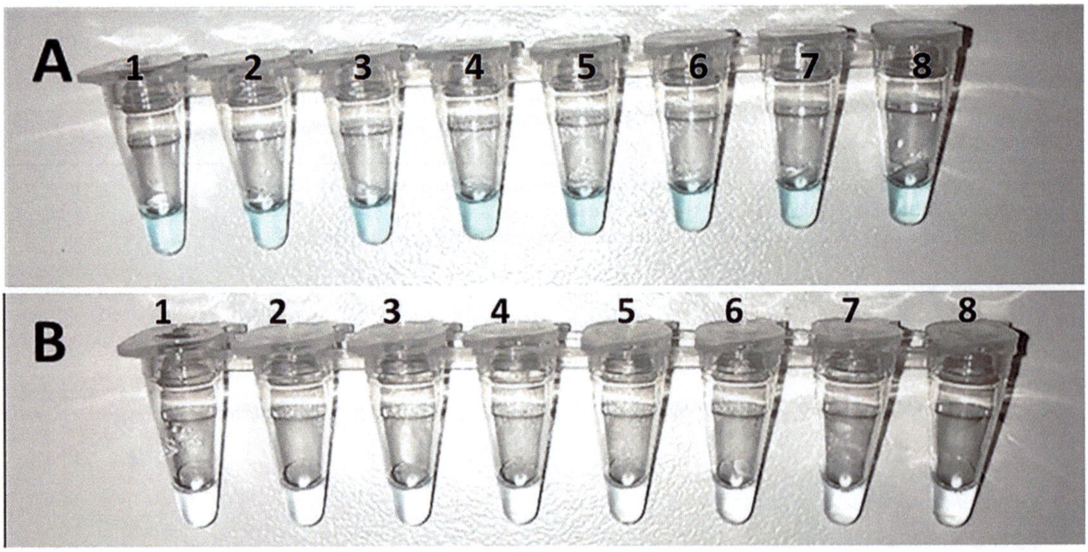

Fig. 2 (**a**) Positive samples for fungal contamination turning light blue/turquoise. (**b**) Negative samples resulting in colorless tubes

4 Notes

1. PBS is intended for cleaning of the sample; consequently, any other general buffer/solution may be used such as 0.9% NaCl solution or TE.
2. Other colorimetric Master Mixes may be used such as WarmStart® Colorimetric LAMP 2× Master Mix (DNA & RNA) (New England BioLabs Inc., Ipswich, Massachusetts, United States). Furthermore, a fluorescence metal indicator like calcein, hydroxy naphthol blue (HNB), or malachite green may be used.
3. Any heating device, such as a water bath or heating block, with the capability of reaching between 60 and 65 °C may be used.
4. The incubation temperature depends on the selected filamentous fungi to be detected. For example, species of *Penicillium* are typically incubated at 25 °C, while species of the heat-resistant *Neosartorya* are incubated at 30 °C.
5. Always include non-spiked samples serving as negative controls to ensure that the samples are not previously contaminated/ spoiled.
6. LAMP primers may be designed with different software. In this case, Primer Explorer V4 (http://primerexplorer.jp/elamp4.0.0/index.html) was selected since it is free for use. For the design process, the sequence of the target gene can serve as input, obtained from sources such as NCBI's RefSeq. Alternatively, multiple sequences from the same database can be retrieved, aligned, and the consensus sequence of a conserved region may be employed. Free tools for visualization and alignment, such as CLC Sequence Viewer [22], are also available for this purpose. Before acquiring newly designed primers, it is essential to assess their inclusivity and exclusivity through in silico testing, both initially and later through in vitro testing. The initial phase can be conducted via a BLAST search (https://blast.ncbi.nlm.nih.gov/Blast.cgi). In the in vitro phase, the optimized assay should be applied to a panel of both target and nontarget strains. The optimal primers and probes concentration must be optimized for optimal performance. In the case of LAMP, the first step of the optimization consists of determining the optimal ratio of outer and inner primers. For this purpose, the concentration of the F3/B3 primers is kept constant at 200 nM and different ratios are tested (e.g., 1:4, 1:5, 1:6 of outer/inner primers). In continuation, the loop primer concentration is optimized by testing concentrations ranging from 300 up to 600 nM.

References

1. Loureiro V (2000) Spoilage yeasts in foods and beverages. Food Res Int 33:247–256
2. García T, Mayoral B, González I et al (2004) Enumeration of yeasts in dairy products: a comparison of immunological and genetic techniques. J Food Prot 67:357–364
3. Roumani F, Azinheiro S, Rodrigues C et al (2022) Development of a real-time PCR assay with an internal amplification control for the detection of spoilage fungi in fruit preparations. Food Control 135:108783
4. Luque MI, Rodríguez A, Andrade MJ et al (2012) Development of a PCR protocol to detect aflatoxigenic molds in food products. J Food Prot 75:85–94
5. Rodríguez A, Rodríguez M, Andrade MJ et al (2012) Development of a multiplex real-time PCR to quantify aflatoxin, ochratoxin A and patulin producing molds in foods. Int J Food Microbiol 155:10–18
6. Rawsthorne H, Phister TG (2006) A real-time PCR assay for the enumeration and detection of *Zygosaccharomyces bailii* from wine and fruit juices. Int J Food Microbiol 112:1–7
7. Cremonesi P, Garofalo C, Picozzi C et al (2022) Development of quantitative real-time PCR and digital droplet-PCR assays for rapid and early detection of the spoilage yeasts *Saccharomycopsis fibuligera* and *Wickerhamomyces anomalus* in bread. Food Microbiol 101:103894
8. Notomi T, Okayama H, Masubuchi H et al (2000) Loop-mediated isothermal amplification of DNA. Nucleic Acids Res 28:63
9. Frisch LM, Mann MA, Marek DN et al (2021) Development and optimization of a Loop-mediated isothermal amplification (LAMP) assay for the species-specific detection of *Penicillium expansum*. Food Microbiol 95:103681
10. Hayashi N, Arai R, Tada S et al (2007) Detection and identification of *Brettanomyces/Dekkera* sp. yeasts with a loop-mediated isothermal amplification method. Food Microbiol 24:778–785
11. Kaneko H, Kawana T, Fukushima E et al (2007) Tolerance of loop-mediated isothermal amplification to a culture medium and biological substances. J Biochem Biophys Methods 70:499–501
12. Roumani F, Azinheiro S, Carvalho J et al (2021) Loop-mediated isothermal amplification combined with immunomagnetic separation and propidium monoazide for the specific detection of viable *Listeria monocytogenes* in milk products, with an internal amplification control. Food Control 125:107975
13. Garrido-Maestu A, Azinheiro S, Carvalho J et al (2017) Combination of microfluidic loop-mediated isothermal amplification with gold nanoparticles for rapid detection of *Salmonella* spp. in food samples. Front Microbiol 8:2159
14. Zhang M, Liu J, Shen Z et al (2021) A newly developed paper embedded microchip based on LAMP for rapid multiple detections of foodborne pathogens. BMC Microbiol 21:1–13
15. Roumani F, Azinheiro S, Sousa H et al (2021) Optimization and clinical evaluation of a multi-target Loop-mediated isothermal amplification assay for the detection of SARS-CoV-2 in nasopharyngeal samples. Viruses 13:940
16. Jeon SB, Seo DJ, Oh H et al (2017) Development of one-step reverse transcription Loop-mediated isothermal amplification for norovirus detection in oysters. Food Control 73:1002–1009
17. Wani LA, Jawa P, Khan JA (2023) Development of one step colorimetric RT-LAMP assays for rapid detection of Apple mosaic virus and Prunus necrotic ringspot virus. J Virol Methods 316:114729
18. Niessen L, Bechtner J, Fodil S et al (2018) LAMP-based group specific detection of aflatoxin producers within *Aspergillus* section *Flavi* in food raw materials, spices, and dried fruit using neutral red for visible-light signal detection. Int J Food Microbiol 266:241–250
19. Ferrara M, Logrieco AF, Moretti A et al (2020) A Loop-mediated isothermal amplification (LAMP) assay for rapid detection of fumonisin-producing *Aspergillus* species. Food Microbiol 90:103469
20. Zhang WW, Jin Y, Ye ZM et al (2015) Development of Loop-mediated isothermal amplification assay for rapid screening of fungal contamination in pepper and paprika powder. Qual Assur Saf Crop Foods 7:97–102
21. Nakayama T, Yamazaki T, Yo A et al (2017) Detection of fungi from an indoor environment using Loop-mediated isothermal amplification (LAMP) method. Biocontrol Sci 22:97–104
22. Bio-Qiagen CLC (2016) CLC sequence viewer
23. Roumani F, Gómez S, Rodrigues C et al (2022) Development and evaluation of a real-time fluorescence, and naked-eye colorimetric, Loop-mediated isothermal amplification-based method for the rapid detection of spoilage fungi in fruit preparations. Food Control 135:108784

Chapter 15

Detection of Aflatoxin-Producing Fungi in Food Commodities Using a Neutral Red-Based Visual LAMP Assay

Ludwig Niessen, Julia D. Bechtner, and Marta H. Taniwaki

Abstract

Aflatoxins are a group of mycotoxins that pose risks to human and animal health due to their high toxicity and carcinogenicity. They are produced by 28 different species within sections *Flavi* (22 species), *Ochraceorosei* (2), and *Nidulantes* (4) of the fungal genus *Aspergillus* (*A*). Rapid detection of aflatoxin-producing fungal species may aid the food industry to assess potential risks associated with commodities. Species-specific LAMP assays were previously developed for the detection of the most common aflatoxin producers, i.e., *A. flavus* and *A. parasiticus*. However, these assays are too specific and therefore fail to detect species that occur less frequently. This chapter describes the development of a group-specific LAMP assay targeting the *aflD* (= *nor1*) gene in all aflatoxin-producing species within section *Flavi* of the genus *Aspergillus*. Detection of positive LAMP reactions under daylight conditions was facilitated using neutral red as visual indirect in-tube signal indicator. The assay was characterized for specificity and sensitivity using purified genomic DNA. A protocol involving minimal sample preparation was developed and applied to the analysis of pure culture spore solutions and to the detection of aflatoxinogenic fungi in naturally contaminated commodities with rice given as an example.

Key words Isothermal amplification, Aflatoxin, *Aspergillus*, Section *Flavi*, Group-specific detection, Screening, Rice

1 Introduction

Aflatoxins are a group of mycotoxins exhibiting acute immuno- and hepatotoxicity in humans and animals [1, 2]. In addition, they have a very high carcinogenic potential following chronic exposure [3]. Due to their pronounced adverse health effects, aflatoxins are an imminent threat for the safety of food and feed, worldwide [4]. Many countries in the world, including the United States and the European Community, have established strict thresholds for aflatoxin concentrations in certain food commodities [5, 6]. In addition to maize, aflatoxins were found to contaminate several other food commodities mainly under warm and dry climatic conditions including cereals, rice, sorghum, pearl millet, peanuts and

Alejandro Garrido-Maestu and Alexandre Lamas (eds.), *Foodborne Pathogens: Loop-mediated Isothermal Amplification Protocols*, Methods and Protocols in Food Science, https://doi.org/10.1007/978-1-0716-4526-0_15,

other nuts and edible seeds, dried fruits, spices and herbs, green coffee beans, tea, and cocoa beans [6]. Aflatoxins can be produced at all stages of commodity processing including growth in the field, harvest, processing, transportation, and storage [7, 8]. Aflatoxigenic species occur in three sections of the fungal genus *Aspergillus*, i.e., sections *Flavi*, *Nidulantes*, and *Ochraceorosei* [9–11]. Among these, section *Flavi* currently contains 22 species capable of producing a wide array of toxic compounds [10–12], the most common of which are aflatoxins type B (AFLB) and type G (AFLG), cyclopiazonic acid (CPA), as well as aspergillic acid and kojic acid. From an economic and toxicological point of view, *A. flavus* and *A. parasiticus* are the most important aflatoxigenic species [13]. In addition, *A. nomius* and *A. pseudonomius* were found to be highly prevalent as producers of aflatoxins in Brazil nuts [14, 15] and also in cocoa [16]. Apart from the established set of three major aflatoxigenic species, a variety of less frequent taxa within *Aspergillus* section *Flavi* have been described as producers of aflatoxins [12, 17]. Moreover, *A. ochraceoroseus* [18] and *A. rambelii* [19] are aflatoxinogenic species in section *Ochraceorosei*, and in section *Nidulantes* (formermy *Emericalla*), *Aspergillus astellata* [20], *A. venezuelensis* [21], *A. olivicola* [22], and *A. miraensis* were shown to be producers of aflatoxins and other potent mycotoxins [23]. The latter six species can produce B-type aflatoxins, but not G type. The presence of any of the abovementioned species in food commodities or products may indicate the potential presence of aflatoxins.

Therefore, knowledge about the presence or absence of aflatoxinogenic molds in food and food commodities can be an important indicator for food safety as well as for aflatoxins being potentially produced later during storage and processing [24, 25]. Classical methods of identification and detection of aflatoxin-producing fungi in food and food raw materials involve time-consuming and costly pure culturing that is followed by cultivation of isolates on different agar media at different temperatures before micro- and macro-morphological analysis by highly spcialized personnel is applied [26]. As an alternative, molecular methods such as PCR and real-time PCR have been applied to the diagnosis of species in *Aspergillus* by amplifying housekeeping genes such as *btub* [27, 28], genes coding for ribosomal RNA [29–31], or, most commonly, genes involved in the biosynthesis of aflatoxins [27, 28, 32–41]. However, the requirement for trained personnel as well as specialized reagents and equipment hampers the application of PCR-based assays at the point of analysis in the industry or in the field. In addition, PCR-based methods need highly purified and concentrated DNA as template since the complex mixture of inhibiting compounds present in natural sample materials may strongly affect the efficiency and sensitivity of a PCR assay [42, 43].

Loop-mediated isothermal amplification (LAMP) [44] has been described as an easy and rapid alternative method of sequence-dependent DNA amplification that has been successfully used for the diagnosis of bacteria [45], viruses [46], and fungi [47] but also for a variety of protists as well as animal [48] and plant species [49]. In addition to the high specificity and sensitivity of LAMP assays, they are relatively unaffected by inhibitors from growth media or from the analyzed sample matrix [50]. They can be performed with a minimum of lab equipment and have therefore a high potential as analytical tool in the food industry and in the field.

A further advantage of LAMP over PCR-based assays is the use of in-tube detection of a direct or indirect signal that indicates the presence of DNA amplification in LAMP reactions. Such alternative detection methods make agarose gel-based product analysis superfluous and reduce time to result substantially. Recently, the application of complexometric indicators such as hydroxynapthol blue (HNB) [51] or various different pH indicators [52] has opened up the possibility of naked-eye in-tube signal detection under daylight conditions that allows to run and interpret LAMP reactions outside of a fully equipped laboratory and might therefore pave the way for field applicable molecular diagnostic assays and their application in agriculture and in the food industry.

We describe here the development of a simple and rapid LAMP-based assay for the group-specific detection of aflatoxin-producing species within *Aspergillus* section *Flavi* using neutral red as a visual indicator for positive LAMP reactions under daylight conditions. Moreover, we describe the application of this assay for the detection of target fungi from purified DNA as well as directly from spore suspensions and naturally contaminated rice after minimal sample preparation.

2 Materials

All buffers and solutions are prepared with ultrapure deinonized water and are filter sterilized through a discardable sterile filter with 0.2 μm pore size and stored as described below. Separate sets of pipettes must be used for manipulations involving DNA and DNA-free manipulations. All pipetting must be done with sterilized single-use filter pipette tips to prevent DNA carryover. All plasticware must be certified to be free of DNA, RNA, and nucleic acid-degrading enzymes. Preparation of LAMP master mixes must be done in a lab area remote from DNA preparation and LAMP reaction incubation/result documentation (*see* **Note 1**).

2.1 Microbial Cultures

Obtain a collection of fungal, yeast, and bacterial pure cultures that is representative of the set of species which the LAMP assay will be supposed to detect. In order to check for negative control reactions, also include species other than the target species that may potentially occur on sample substrates (*see* **Note 2**). Also include the sample matrix or matrices to which the LAMP assay is supposed to be applied as negative controls.

2.2 Culture Media

Prepare the following nutrient media for cultivation of fungal mycelia.

1. Malt extract broth, 2%: weigh 20 g malt extract and 2 g soy peptone. Add tap water to 1 L and adjust pH to 5.4. Mix well and autoclave at 121 °C for 20 min.
2. Malt extract agar, 2%: weigh components as described under 1, and add 15 g agar (Difco). Add tap water to 1 L and adjust pH to 5.4 before autoclaving at 121 °C for 20 min. Let the medium cool down to 60 °C and fill ca. 20 mL each to 9-cm-diameter sterile Petri dishes. Let the medium dry for 24 h at ambient temperature and store at 4 °C in the dark.

2.3 TAE Buffer

Prepare the following solution for 1 L of a 50× stock solution

1. Weigh 242.28 g of Tris-base, 60.02 g 100% acetic acid, and 37.22 g EDTA disodium salt.
2. Dissolve Tris base in 700 mL distilled water.
3. Carefully add 60.02 g of acetic acid and 37.22 g EDTA disodium salt.
4. Add distilled water to 1 L and make sure all components have dissolved.
5. Adjust pH to 8.5.
6. For a 1× TAE buffer, fill 20 mL of 50× TAE stock to 1 L with distilled water and mix well.

2.4 LAMP Master Mix

Prepare the following solutions

1. 10× LAMP buffer: 100 mM $(NH_4)_2SO_4$, 100 mM KCl, pH 8.7 (*see* **Note 3**). All components should be analytical grade.

   To prepare 50 mL of 10× LAMP buffer, weigh 660.7 mg of $(NH_4)_2SO_4$, 372.7 mg of KCl and add ultrapure water to 40 mL. Stir until all reagents have dissolved. Adjust pH to 8.7 with 1 M NaOH solution. Add buffer solution to 50 mL volumetric flask and add ultrapure water to final volume. Mix well and filter sterilize solution through 0.2 μm filter cartridge into sterile 50 mL Falcon tube. Store at 4 °C or freeze at −20 °C in aliquotes.

2. $MgCl_2$ stock solution: 200 mM $MgCl_2$.

   Weigh 2.033 g of $MgCl_2 \times 6\ H_2O$ and add ultrapure water to 40 mL. Stirr until dissolved. Transfer solution to 50 mL volumetric flask and fill to final volume with ultrapure water. Mix well and filter sterilize solution through 0.2 μm filter cartridge into sterile 50 mL Falcon tube. Store at 4 °C or freeze at −20 °C in aliquots.

3. Neutral red indicator solution: 2.5 mM in 1× LAMP buffer, pH 8.7.

   Weigh 10.8 mg of neutral red (3-amino-7-dimethylamino-2-methylphenazin hydrochloride, Sigma Aldrich) and add 15 mL of 1× LAMP buffer (dilute 10× LAMP buffer 1:10 in ultrapure water and filter sterilize). Store solution at 4 °C (*see* **Note 4**)

4. Nucleotide mix: 10 mM each G, A, T, C.

   Mix 1 mL each of 100 mM stock solutions of the desoxynucleoside triphosphates dGTP, dATP, dTTP, and dCTP and add 6 mL of ultrapure water. Mix well and prepare aliquots of 500 μL in sterile 1.5 mL reaction tubes. Store at −20 °C for long-term storage. Store aliquots at 4 °C after thawing.

5. *Bst* DNA polymerase, large fragment (*see* **Note 5**).

   Make aliquots of 20 μL from 8000 U/mL stock and freeze immediately at −20 °C.

6. LAMP primer mix: Solution for 80 reactions.

   Order oligonucleotide primers in freeze dried form. Add sterile ultrapure water to prepare 50 μM stock solutions (see data sheet from supplier). Mix 80 μL each of primers FIP-norl ID8 and BIP-norl ID8 (see Table 1), 40 μL each of primers

Table 1
Primers used in the LAMP assay for the detection of aflatoxin-producing *Aspergillus* spp

Primer name	Sequence 5′ > 3′	References
F3-norl ID8	ACT GCG ACT CGG AAA CYG A	[58]
B3-norl ID8	GGA CTG CTG CAG CAT CAG	[58]
FIP-norl ID8	GGC CCA AAG TTC GTC GCC AT-C CAG ACA TTG CGG GAR GA[a]	[58]
BIP-norl ID8	ACC ATG CCC CTC GAR CAT CT-G CGG GTT GCC TGA AAC AG[a]	[58]
LF-norl ID92	ACY ACC ACG TCC AAG TGC	[58]
LB-norl ID92	TGA TGG TCA AYA TGT ATG CTC C	[58]
F2-norl ID8	CCA GAC ATT GCG GGA RGA	
B2-norl ID8	GCG GGT TGC CTG AAA CAG	

Modified from Niessen et al. [58]
[a]Hyphen indicates junction between F1c/B1c (left) and F2/B2 (right) parts of FIB/BIP primers.

LF-norl ID92 and LB-norl ID92, and 10 μL each of primers F3-norl ID8 and B3-norl ID8. Mix well and store at 4 °C. Do not prepare primer mix for more than 80 reactions in advance.

2.5 Other Components

1. Disposible plasticware: 0.2 μm sterile filter cartridge; 200 μL pyrogen-free PCR tubes with flat lid; sterile 500 μL and 1.5 mL reaction tubes.
2. Three sets of pipettes including 200–1000, 50–200, 20–100, 2–20, and 1–10 μL pipetting volume (one set for work including DNA, one set for DNA-free work in the general lab area, one set for DNA-free work in the "Master Mix Preparation" area of the lab).
3. Gradient thermal cycler.
4. Digital camera or other documentation system for color prints.
5. Ultrapure water or double distilled water.

3 Methods

3.1 Selection of a Group-Specific Target Sequence (See Note 6)

1. Find the complete aflatoxin pathway gene cluster of *Aspergillus parasiticus* NRRL 5862 in GenBank under accession no. AY371490. Use your browser's search function to search the site for the textstring *gene* = *"aflD"*. Hit the "gene" link to see the gene sequence marked within the total accession sequence. Hit the "display as FASTA" link at the bottom of the page to show the *aflD* gene sequence only (nucleotides 17,448 to 18,437). The sequence should be 990 nt long. Paste the sequence to a sequence alignment editor software such as BioEdit (*see* **Note** 7).
2. Open the BLAST function of the NCBI database under https://blast.ncbi.nlm.nih.gov/Blast.cgi and enter the name of a typical aflatoxin-producing *Aspergillus* species in the "BLAST Genomes" field and select the species name from the selection list. Press "Search."
3. Paste the *A. parasiticus* NRRL 5862 *aflD* gene sequence from **step 1** into the field marked with "Enter Query Sequence" and make sure that "Highly similar sequences (megablast)" is selected in the "Program Selection" section. Press "BLAST" (*see* **Note 8**).
4. Select the sequence accession with 100% sequence coverage and maximum percentage identity and hit the link under "Description" to see an alignment of the query sequence with the identified subject sequence.
5. Hit the "GenBank" link given with the range description of the subject sequence to see the selected gene locus in GenBank

format. Hit the "FASTA" link to see the sequence in FASTA format. Copy the sequence descriptor including the ">" and the complete sequence and paste it to a sequence alignment editor software such as BioEdit (*see* **Note 7**).

6. Repeat the procedure with as many typically aflatoxigenic species (as shown in [17]) as possible and align all sequences using the ClustallW multiple alignment tool available in most sequence alignment editors. Figure 1 shows a multiple alignment of a partial sequence (209 nt) of the *nor1* genes of typical aflatoxin-producing *Aspergillus* spp. belonging to section *Flavi*.
7. Use the contig assembly function of your sequence alignment editor to create a consensus sequence of the target gene.

3.2 Design of Group-Specific LAMP Primers for Aflatoxin Producers

1. Copy the consensus sequence generated in Subheading 3.1, **step** 7 into a word processor and make sure to delete all inserts ("-") from the text. The consensus sequence should now be 990 nt in length. Copy nt 293 to 522 from the consensus sequence and paste to a new document of your text editor. Safe sequence as plain text file (.txt) using the DOS formatting option.
2. Open the PrimerExplorer V5 software tool provided by Eiken Chemicals Co. Ltd. (Tokyo, Japan) under the link https://primerexplorer.jp/e/ and follow instructions for the design of regular primers: click "search" under "Target Sequence File/Primer Information File" and browse your computer for the .txt-file previously generated under step 1 to serve as target sequence. Use "automatic judgement" as your parameter set for primer design and press "Primer Design" to upload the target sequence into the program.
3. Press the "Generate" button to generate sets of LAMP primers (*see* **Note 9**). Up to five sets will be loaded which the program determines to be optimum in regard to primer length, melting temperature, and secondary structure. To view the primers within the target sequence, press "Display." To see all generated primer sets, or to change parameters for primer design, press the "Detail Settings" button. To see all primer sets generated with the modified conditions, press "Generate" again and then "Display." Up to 1000 primer sets will be displayed in a new window as the "Primer Set List."
4. Mark the primer sets of interest in the lefthand checkbox and press the "Confirm" button. All chosen primer sets will be displayed in a new window as the "PrimerSet Confirm page" (*see* **Note 9**).

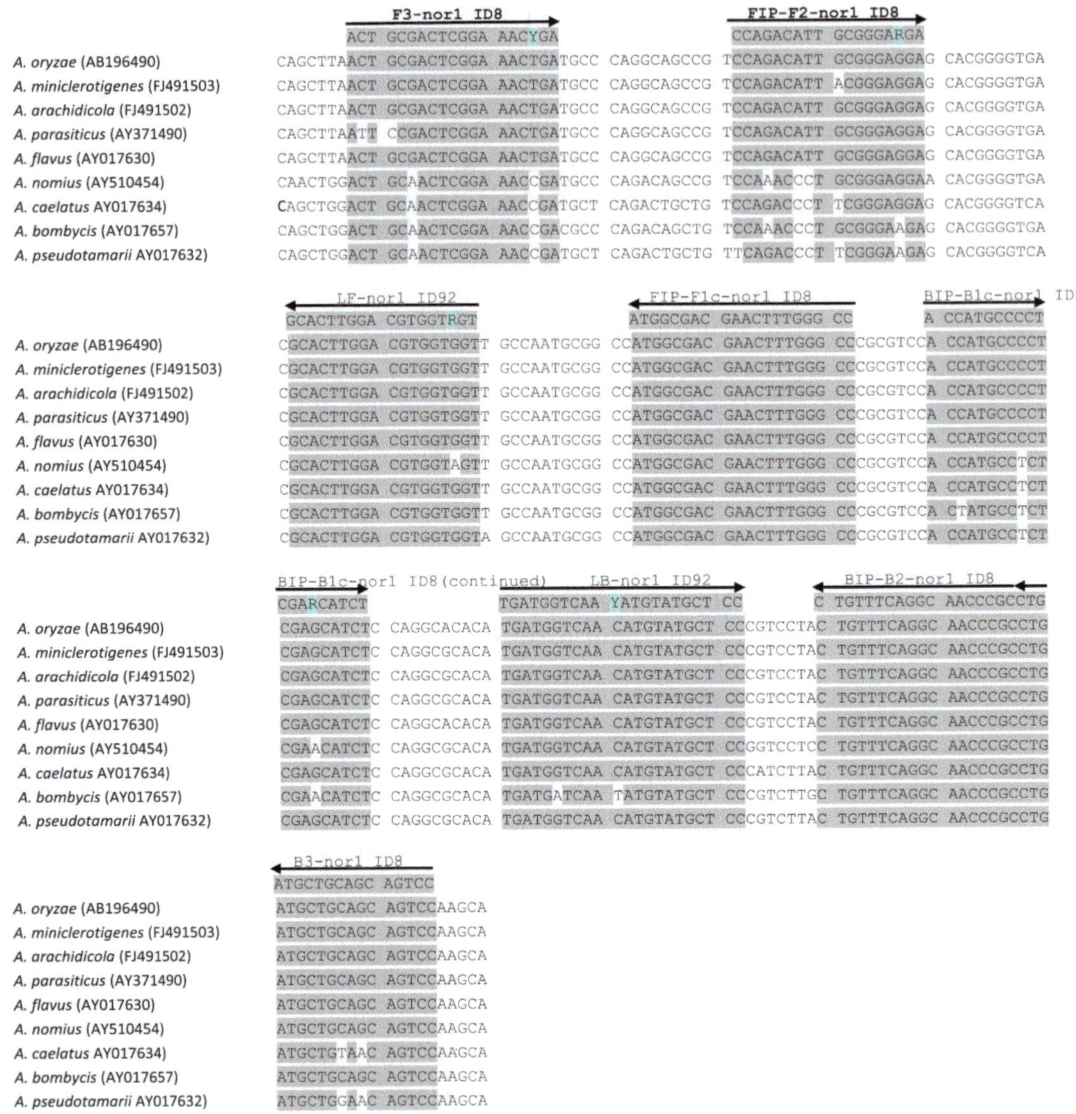

Fig. 1 Multiple alignment of a partial sequence (209 nt) of the *nor1* genes of typical aflatoxin-producing *Aspergillus* spp. belonging to section *Flavi.* Positions of nor1-specific LAMP primers (F3, F2, F1c, B3, B2, B1c) and a pair of loop primers (loopF and loopB) are marked in gray. Positions marked in turquoise are wobble positions in the respective primer. Arrowheads indicate orientation of the primers used (=> forward primer, <= reverse primer)

5. Use the BLAST algorithm under http://www.blast.ncbi.nlm.nih.gov/ to search GenBank for off-target matches of each primer from the "PrimerSet Confirm page" with nontarget organisms to check the in silico specificity. Make sure to check the F2/B2 and F1c/B1c parts of primers FIP/BIP in separate BLAST searches. Ideally, all primers should show high-score hits exclusively with the target gene in several typical aflatoxin producers within *Aspergillus.* Low-score hits with a variety of

nontarget species will occur regularly and can be ignored (*see* **Note 10**).

6. If high-score BLAST hits occur with nontarget sequences, repeat step 5 with the primer set next closely to dG = −2.0 (see also **Note 10** for exceptions).
7. When appropriate primer set or sets have been chosen, press the "Primer Information" button displayed on the "PrimerSet Confirm" page to create a PrimerInfo file in plain text format. Save the PrimerInfo file for later use. Load the target sequence into a word processor and mark the positions of each selected primer for later publication and personal information.
8. Use the loop primer design option of the PrimerExplorer software to create additional primers (loop primers) in order to accelerate the amplification reaction (*see* **Note 11**). Start PrimerExplorer and use the PrimerInfo file previously saved as target sequence for loop primer design. Press "Primer Design" to see the target sequence with the previously designed LAMP primers at their binding sites and repeat **steps 3–7** to design loop primers. Figure 1 shows the LAMP primers and loop primers given in Table 1 as arrows in the multiple alignment of the *nor1* gene with non-matching nucleotides marked in white.
9. Order LAMP primers and loop primers as displayed in the green shaded part of the primer set on the "PrimerSet Confirm" page. In case of manual loop primer design, make sure LoopF primer is reverse complement, whereas LoopB primer is according with the target sequence.
10. Table 1 lists the primers that were selected based on the *aflD* gene of *Aspergillus* spp. for the group-specific detection of typical aflatoxin producers.

3.3 Preparation of LAMP Master Mix (See Note 12)

1. For a 25 μL reaction volume, mix the following components in a sterile reaction tube (500 μL–2 mL) using the DNA-free pipette set in the dedicated "Master Mix Preparation Area" of your lab: 2.5 μL of 10× LAMP buffer (see Subheading 2.4, **step 1**), 1 μL $MgCl_2$ stock solution (Subheading 2.4, **step 2**), 3.5 μL nucleotide mix (Subheading 2.4, **step 4**), 2.6 μL primer mix (Subheading 2.4, **step 6**), 8.4 μL filter-sterilized ultrapure DNA free water, 1 μL *Bst* DNA polymerase, large fragment from 8000 U/mL stock (Subheading 2.4, **step 5**), and 1 μL neutral red indicator solution (Subheading 2.4, **step 3**), and mix well.
2. Use the DNA-free pipette set (Subheading 2.5, **step 2**) and pipette tips with filter to distribute 20 μL per reaction of the master mix to sperile and pyrogen-free 200 μL PCR reaction tubes with tightly fitting flat lid.
3. Close lids immediately and make sure the master mix has a yellow to orange color.

3.4 Assay Set Up: Temperature Optimization

1. Prepare in sterile ultrapure DNA-free water a solution of 100 ng/μL of highly purified genomic DNA from a strain of a typical aflatoxin-producing *Aspergillus* spp., e.g., *Aspergillus parsiticus* CBS 126.62 (*see* **Note 13**).
2. Prepare master mix for 13 reactions as described in Subheading 3.3, **step 1** and add per reaction 5 μL of the DNA dilution prepared in **step 1**. Make sure the color of the master mix is still yellow.
3. Mix well and distribute 25 μL per reaction to 200 μL PCR reaction tubes with flat lid and close lids tightly.
4. Use a gradient thermal cycler and set the cycling program as follows: one cycle of 90 min with a temperature gradient set from 61 °C to 71 °C in 12 increments so that each column of the heating block has a different constant temperature (*see* **Note 14**).
5. Select the highest temperature at which a color change from yellow to pink has occurred after 90 min of incubation as the incubation temperature for further experiments (67.9 °C; see Fig. 2).

3.5 LAMP Product Verification

1. Separate the amplicons of a complete LAMP reaction on a 1.3% agarose gel prepared in 1× TAE buffer (see Subheading 2.3) at 120 V/100 mA together with a 1 Kbp DNA size marker and stain gel with dimidium bromide or other DNA gel stain.
2. Cut the smallest LAMP fragment (ca. 260 bp) from the gel and extract DNA using a gel extraction kit according to the manufacturer's recommendations.
3. Order the F2 and B2 parts of primers FIP and BIP (see Table 1) as separate primers and prepare seven PCR reactions plus a negative control using 1 μL of purified LAMP product per reaction as template and primers F2-norl ID8 and B2-norl ID8.

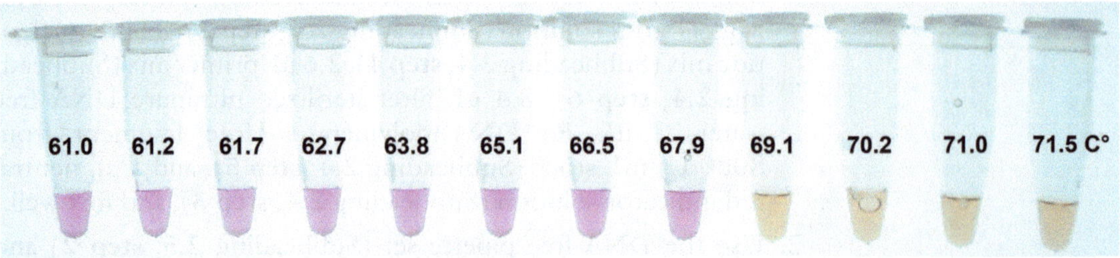

Fig. 2 Optimization of incubation temperature. LAMP reactions run in a gradient thermal cycler with primerset nor1 ID8/ID92 and genomic DNA of *A. parasiticus* CBS 126.62 (500 ng/reaction) after 90 min incubation. Numbers indicate the incubation temperature of individual reactions. Pink color indicates positive LAMP reaction due to in vitro DNA biosynthesis followed by a pH drop

4. Run PCR under the following cycling program: 1 cycle initial denaturation at 95 °C for 5 min, 35 cycles of 1 min denaturation at 95 °C, 45 s annealing at 57.3 °C, 15 s extention at 72 °C, and 1 cycle of 5 min final extention at 72 °C.
5. Separate all reactions on a 1.3% agarose gel in 1× TAE buffer at 100 V/90 mA with a 1 Kbp size marker and stain.
6. Excise all PCR fragments of ca. 260 bp and extract from gel using a gel DNA extraction kit according to the manufacturer's recommendations.
7. Obtain the nucleotide sequence of the PCR product by bidirectional Sanger sequencing using primers F2-norl ID8 and B2-norl ID8 in separate reactions.
8. Use both nucleotide sequences to obtain a consensus sequence and compare with the *aflD* sequence of *A. parasiticus* CBS 126.62 (accession no. AY371490) or other *A. parasiticus* strain.
9. The sequence covered by the consensus sequence should show at least 90% identity with the *aflD* gene's nucleotide sequence (*see* **Note 15**).

3.6 Assay Setup: Sensitivity Testing

1. Prepare a tenfold serial dilution of the purified gDNA of known concentration (see Subheading 3.4, **step 1**) in sterile ultrapure water starting from 10^{-1} to 10^{-7}.
2. Prepare a master mix for nine reactions as described in Subheading 3.3, **steps 1–3** and distribute 20 μL each to eight 200 μL PCR caps and close lids.
3. Add 5 μL from each serial dilution to master mixes 1–7 and add 5 μL of sterile ultrapure water to master mix 8 as no template control (NTC).
4. Incubate in a heating block at optimized temperature (see Subheading 3.4, **step 5**).
5. Figure 3 shows LAMP reactions after 60 min of incubation at 67 °C with a serial dilution of purified *A. parasiticus* CBS 126.62 gDNA as template as visually inspected at daylight conditions (*see* **Note 16**). The assay has a detection limit of 11 pg/rxn for *A. parasiticus* genomic DNA (*see* **Note 17**).

3.7 Assay Setup: Specificity Testing

1. Prepare purified genomic DNA from reference strains of the most closely related fungal species (*see* **Note 13**). We have analyzed 42 different species of *Aspergillus* and 3 species of *Emericella* as listed in Table 2 (*see* **Note 18**).
2. Dilute genomic DNA to a concentration of 100 ng/μL in sterile ultrapure water. If assay is planned to be used for analysis of sample materials, prepare also DNA from non-contaminated

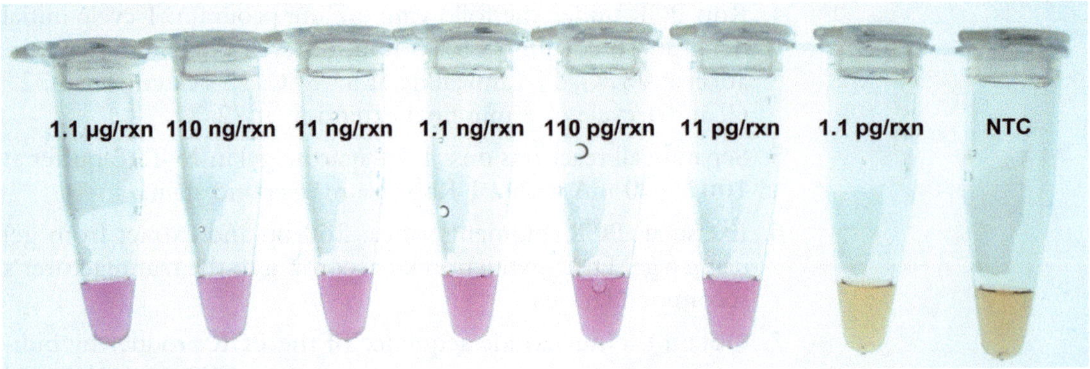

Fig. 3 Sensitivity of the *aflD*-specific LAMP assay with neutral red for signal detection and a tenfold serial dilution of gDNA from *A. parasiticus* CBS 126.62 as template after 60 min incubation at 67 °C. Visual inspection under day light conditions. Pink, positive reaction; yellow, negative reaction; NTC, no template control with sterile ultrapure water instead of template. Numbers show template DNA concentrations per reaction (rxn)

sample matrix or matrices in order to rule out cross reactions with uncontaminated sample materials.

3. Prepare a dilution of *A. parasiticus* CBS 126.62 DNA with a concentration near to the detection limit of the assay, e.g., 100 pg/µL, in sterile ultrapure water as positive control.
4. Prepare master mix for the desired number of reactions including positive and negative controls and distribute 20 µL per PCR tube. Close lids immediately.
5. Add 5 µL of diluted DNA for each fungal strain and for the positive control. Add 5 µL of sterile ultrapure water as negative control. Incubate for 60 min at 67 °C.
6. Remove reactions from heating block and inspect visually under daylight conditions. A color change from yellow to pink should occur in 15 of the 45 samples listed in Table 2.
7. Compare results with literature data about aflatoxin producing abilities of the tested fungal species (*see* **Note 19**). In cases in which positive LAMP results should occur with DNA of non-aflatoxigenic species, the assay conditions need to be adjusted for higher specificity.

3.8 Assay Setup: Specificity adjustment

1. Run the LAMP assay with DNA of those species that showed false-positive results under the coditions described under Subheader 3.7, including positive and negative controls. Increase incubation temperature by 0.5 °C. Repeat until none of the false positives show a positive result.
2. If the strategy described in **step 1** does not solve the problem, use formamide as additive to enhance specificity. Prepare for each false-positive template DNA and for the reference DNA a

Table 2
LAMP-based assay for potential aflatoxin producers. Specificity testing using gDNA purified from pure cultures of 45 *Aspergillus* spp. and three *Emericella* species

Genus	Species	Strain	Clone	LAMP result
Aspergillus	*aculeatus*	TMW 4.1776[a]	TMW 4.1776	–
	aflatoxiformans	IBT 3850[b]	TMW 4.2206	+
	alliaceus	*DSM 8133*[c]	*TMW 4.1077*	–
	arachidicola	IBT 27128	TMW 4.2204	+
	auricomus	CBS 467.65[d]	TMW 4.1631	–
Emericella	*astellata*	IBT 21903	TMW 4.2202	–
	awamori	CBS 101704	TMW 4.1066	–
	bombycis	IBT 23536	TMW 4.2210	+
	bridgeri	CBS 350.81	TMW 4.1632	–
	caelatus	IBT 29700	TMW 4.2209	–
	carbonarius	TMW 4.1512	TMW 4.1512	–
	clavatus	CBS 513.65	TMW 4.1086	–
	elegans	CBS 310.80	TMW 4.1633	–
	ellipticus	CBS 707.79	TMW 4.1629	–
	flavus	TMW 4.1829	TMW 4.1829	+
	foetidus	CBS114.49	TMW 4.1628	–
	fumigatus	CBS 113.55	TMW 4.0623	–
	helicothrix	CBS 677.79	TMW 4.1630	–
	heteromorphus	CBS 117.55	TMW 4.1626	–
	insulicola	CBS 382.75	TMW 4.1634	–
	japonicus	CBS 114.51	TMW 4.1627	–
	mottae	MUM 10.231[e]	TMW4.2791	+
	minisclerotigenes	IBT 27177	TMW 4.2205	+
	niger	CBS 101698	TMW 4.1068	–
	nomiae	CBS 260.88	TMW 4.1960	+
	novoparasiticus	MUM 15.15	TMW 4.2796	+
	ochraceoroseus	CBS 101887	TMW 4.1772	–
	ochraceus	CBS 263.67	TMW 4.0706	–
Emericella	*olivicola*	IBT 26499	TMW 4.2201	–
	oryzae	IBT 28103	TMW 4.2208	+
	parasiticus	CBS 126.62	TMW 4.1768	+

(continued)

Table 2 (continued)

Genus	Species	Strain	Clone	LAMP result
	petrakii	CBS 105.57	TMW 4.1087	–
	pseudotamarii	IBT 21092	TMW 4.2212	+
	rambellii	IBT 14580	TMW 4.2211	–
	sclerotiorum	CBS 549.65	TMW 4.1089	–
	sergii	MUM 10.208	TMW 4.2793	+
	sojae	IBT 21643	TMW 4.2207	+
	sulfureus	CBS 550.65	TMW 4.1067	–
	tamarii	CBS 591.68	TMW 4.1771	–
	terreus	CBS 377.64	TMW 4.1060	–
	toxicarius	CBS 822.72	TMW 4.1766	+
	transmontanensis	MUM 10.214	TMW 4.2795	+
	tubingensis	ITEM 4496[f]	TMW 4.2008	–
	usamii	CBS 101700	TMW 4.1072	–
Emericella	*venezuelensis*	IBT 20956	TMW 4.2203	–

Modified from Niessen et al. [58]

[a]*TMW* Lehrstuhl für Technische Mikrobiologie Weihenstephan, Technical University of Munich, Freising, DE

[b]*IBT* ulture collection of Center for Microbial Biotechnology (CMB), Department of Systems Biology, Technical University of Denmark, Lyngby, DK

[c]*DSM* Deutsche Sammlung von Mikroorganismen und Zellkulturen, Darmstadt, DE

[d]*CBS* Westerdijk Fungal Biodiversity Institute, Utrecht, NL

[e]*MUM* Micoteca da Universidade Minho, University of Minho, Braga, PT

[f]*ITEM* Istituto Tossine e Micotossine da Parassiti Vegetali, CNR, Bari, IT

master mix for nine reactions and add 45 μL of the respective DNA from a 100 ng/μL solution. Prepare for each master mix eight PCR tubes and add 0 to 1.75 μL 100% formamide in 0.25 μL increments (equal to 0–7% formamide per reaction). Add 25 μL of the respective master mix and seal tubes tighly. Incubate at 2 °C below the highest temperature used in **step 1**.

3. Select formamide concentration at which no false positives occur and at which the reference DNA still gives a positive LAMP result.
4. If false-positive results still occur, repeat **step 2** and increase incubation temperature in 0.5 °C increments until no false positives occur.
5. Repeat specificity testing as described under Subheader 3.7 to make sure that results are in accordance with the aflatoxin-producing potential of the tested species.

3.9 Assay Application: Direct Testing of Pure Culture Spore Solutions (See Note 20)

1. Grow fungal cultures on 2% malt extract agar (see Subheading 2.2, **step 2**) at ambient temperature until sporulation occurs.
2. Remove conidia from a sporulating fungal culture with a sterile toothpick and transfer to 300 μL of sterile ultrapure water.
3. Add sterile glass beads (0.4 g of 0.5 mm diameter, 0.2 g of 1.25–1.65 mm diameter) and vortex at maximum speed for 10 min or shake vigorously by hand if no vortex is available.
4. Prepare suitable number of master mixes including positive and negative controls as described under Subheading 3.3, **steps 1–3** and distribute 20 μL each to sterile PCR tubes.
5. Add 5 μL of the vortexed conidial solution per reaction and close lids immediately. Incubate for 60 min at 67 °C.
6. Figure 4 shows the *aflD*-specific LAMP assay after incubation of a serial dilution of *A. parasiticus* CBS 125.62 conidia without glass beads treatment (A) and with glass beads treatment (B).

3.10 Assay Application: Testing of Naturally Contaminated Sample Materials (See Note 21)

1. Weigh 2 g of each sample or a whole fig chopped to small pieces, and transfer into a sterile 15 mL falcon tube.
2. Add 5 mL of sterile ultrapure water containing 1% (v/v) of Tween 20 and shake vigorously by hand for 60 s.
3. Immediately transfer 2 mL of the turbid supernatant to a sterile 2 mL reaction vessel and centrifuged for 2 min at 11,000× *g*.
4. Remove Tween 20 by washing the resulting pellet three times each with 1 mL sterile ultrapure water with intermediate centrifugation as described in **step 3**.
5. Suspend the washed pellet in 500 μL of sterile ultrapure water and add sterile glass beads (0.4 g of 0.5 mm diameter, 0.2 g of 1.25–1.65 mm diameter) and vortex at maximum speed for 10 min or shake vigorously by hand.
6. Prepare LAMP master mix for appropriate number of reactions and distribute 20 μL for each reaction, including positive and negative controls.
7. Add 5 μL of template directly from samples prepared according to **step 5** and close lids and incubate for 60 min at 67 °C or other optimized temperature conditions.
8. Inspect visually with the naked eye for reactions in pink (positive) or yellow (negative). Table 3 shows the results of the analysis of Brazilian rice samples taken at different steps during the rice processing chain and compares them with the aflatoxin levels found in corresponding samples (*see* **Note 22**).

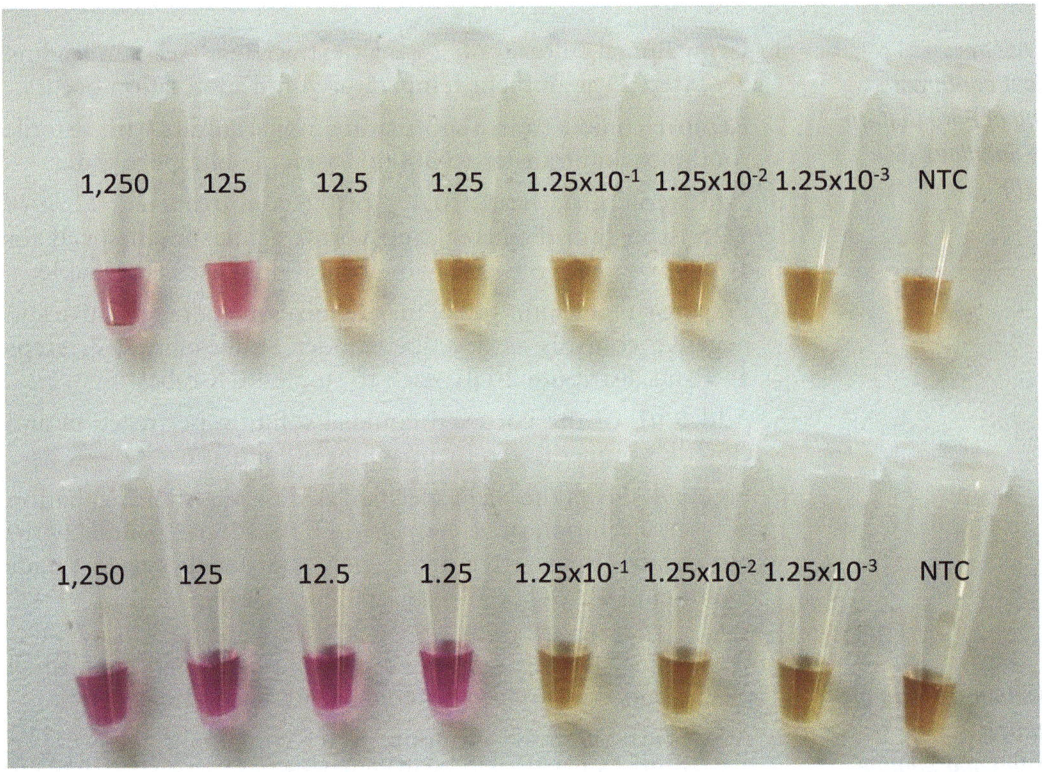

Fig. 4 aflD-specific LAMP assay using a tenfold serial dilution of *A. parasiticus* CBS 126.62 conidia with and without mechanical treatment with glass beads before direct application of the supernatant as template. A. Serial dilution of conidia as template without glass beads treatment prior to addition as template. B. Serial dilution of conidia as template with glass beads treatment prior to addition as template. Unit, conidia per reaction; NTC, no template control with sterile ultrapure water instead of template; pink, positive reaction; yellow, negative reaction

4 Notes

1. LAMP reactions produce high amounts of product DNA within short time. Such high product concentrations tend to result in aerosolized DNA upon opening of reactions vessels after the reaction has terminated. Due to its usually very high sensitivity, the LAMP reaction will amplify aerosolized product DNA when contaminations occur, which will lead to positive reactions in negative controls. Should product contamination occur, the affected lab area including pipettes needs to be thoroughly cleaned with a DNA-degrading agent. All buffers and solutions should be newly prepared. In cases where false results still occur, alternative LAMP primers need to be designed, which do not bind to any sequence within the contaminating product.

Table 3
Results (positive, negative) of the aflD LAMP assay performed with surface washings of rice seeds (2 g per sample) as template after processing according to Subheading 3.10, steps 1–7 and results of total afaltoxin analysis

Sample number	Processing stage	Aflatoxins	LAMP
1	Field	ND[a]	–[b]
3	Field	ND	–
4	Field	ND	–
6	Field	ND	–
10	Storage	ND	–
11	Storage	ND	–
12	Storage	ND	–
35	Market	ND	–
38	Market	ND	–
40	Market	ND	+[c]
44	Market	23.4 ng/g	+
47	Market	<0.2 ng/g	–
57	Market	ND	+
61	Market	70.9 ng/g	+
65	Field	ND	–
66	Field	ND	–
67	Field	ND	–
68	Field	ND	–
70	Field	ND	–
71	Field	ND	–
72	Field	ND	–
73	Field	ND	+
75	Field	ND	–
77	Drying yard	ND	–
78	Processing	ND	+
79	Processing	ND	+
80	Processing	ND	–
82	Processing	0.2 ng/g	+
83	Processing	<0.2 ng/g	+

(continued)

Table 3
(continued)

Sample number	Processing stage	Aflatoxins	LAMP
108	Market	<0.2 ng/g	+
109	Market	0.24 ng/g	+
117	Market	ND	–
118	Market	ND	–

[a]Not detected at LOD 0.04 ng/g
[b]Negative result in LAMP assay
[c]Positive result in LAMP assay

2. Make sure that reference cultures come from culture collections that are officially recognized by the World Federation of Culture Collections (WFCC, https://wfcc.info/membership/memberlist) whereever possible. As an alternative, contact authors who have published original descriptions of new species directly to obtain a specimen culture. Use the growth media recommended by the source collection to cultivate specimens. Make sure by culture appearance and micromorphology that cultures are uncontaminated and that their microscopical features match with typical features of the genus under analysis. Make sure also that the used species name is still legitimate by checking the name in the Mycobank database (https://www.mycobank.org/Advanced_names_search). If the name of your specimen differs from the legitimate name, use the legitimate name and provide the specimen's name as a synomym in brackets.
3. This LAMP buffer contains a comparably low concentration of buffering substance in order to make sure that the pH of the solution can drop below pH 7.4 in reactions, in which DNA amplification takes place. The drop in pH is initiated by massive production of H^+-ions as a by-product of enzymatic DNA biosynthesis. At pH 7.4 the neutral red indicator dye changes its color from yellow to red to indicate a positive LAMP reaction.
4. Neutral red does not dissolve completely under the conditions so that the stock solution will be turbid. However, the volume withdrawn from the stock solution (1 μL per reaction) will completely dissolve in the master mix. Make sure to shake the stock solution well before pipetting to ensure homogeneity.
5. *Bst* DNA polymerase, large fragment is now available from a variety of suppliers. The solution described here relates to the enzyme as supplied by New England Biolabs (product code M0275L) where it comes in a concentration of 8,000 U/mL in

glycerol. When using *Bst* DNA polymerase, large fragment from another supplier, make sure to check concentrations and to adjust the volume added per LAMP reaction if necessary.

6. The approach used in the current study for the design of LAMP primers can be readily generalized for other groups of organisms that have certain features in common and that make them distinct from other organisms. We assumed that all aflatoxin-producing *Aspergillus* species should have a full cluster of 25 genes [53, 54] that are involved in the biosynthesis of this group of toxins and that the presence of such genes will be a unique feature common to all species. Due to the relatively close phylogenetic relationship of those species, we further assumed that the coding sequences of such genes should have a certain degree of similarity between species and that sequence regions should exist that are highly similar and run long enough (200–300 bp) to be fit for LAMP primer design. In order to find appropriate sequence information, we made alignments for all known genes within the aflatoxin biosynthetic pathway in a variety of different typical aflatoxin-producing *Aspergillus* and *Emericella* species in order to find highly conserved genes or gene regions. The gene coding for norsolorinic acid reductase (*nor1, aflD*) showed an exceptionally high degree of sequence conservation between different species and was therefore chosen as the basis for the design of LAMP primers that are specific to the detection of those *Aspergillus* species (nearly all of them in section *Flavi*) that are known as producers of aflatoxins in a group-specific manner.
7. Although we used the free BioEdit software package, there are many other software packages on the marked or freely available for download. Please find a comprehensive list of available sequence alignment software tools under https://en.wikipedia.org/wiki/List_of_sequence_alignment_software.
8. The query sequence will now be searched against the complete genomic sequence of the species under analysis. In most cases, the result will show the chromosome 3 sequence of one or more strains of the species under study, depending on how many genomes of that particular species are available in the database.
9. Each generated primer set consists of six oligonucleotides that can be viewed in an alignment with the target sequence in a new window called "primer set list." The program calculates the mean 3′ (forward primers) or 5′ end (reverse primers) stability for each oligonucleotide and provides average ΔG (called the dG (dimer) value) for each primer set. According to our experience, primer sets with a dG value close to −2.0 are optimum for rapid and sensitive assays. Checking the chosen primer or primers and pressing the "Confirmation" button will

open a new page that shows the chosen primer sets as they should be ordered, marked in green. In additon, the F2/B2 and F1c/B1c parts of primers FIP and BIP are displayed. The primer information site also shows additional info for each primer, i.e., alignment positions in the target sequence, length, melting temperature, 3′ and 5′ stability, and GC rate. The primer information site should be saved in html format for later primer ordering. Moreover, pressing the "Primer Information" button will generate a .txt file that contains the target sequence, the respective primer sequences, and the positions of each primer in the target sequence. This file is needed in case loop primers should be designed with the PrimerExplorer program.

10. It may occur that F3 and/or B3 show high score hits also with nontarget sequences. In cases where all other primers of a set show high in silico specificity for their target sequence, this can be tolerated since primers F3 and B3 do only have a function in the generation of the very first product strand during DNA biosynthesis in LAMP reactions and their binding site will be absent from the LAMP product. Therefore, they do not have an influence on the specificity of an assay in which they are used.

11. The amplification reaction in a LAMP assay will generally run with the set of four primers that have been created with the regular primer design module of PrimerExplorer. However, since loop primers bind to the loop structures that are formed as secondary structures during DNA biosynthesis in LAMP, they can provide a higher number of the basic amplicon that is necessary to initiate the reaction. The application of loop primers can strongly accelerate the reaction speed of a LAMP reaction that can reach its endpoint in about 45 min instead of 90 min without loop primers. LoopF primers (LF) are positioned between the 3′ end of the binding site for primer F2 and the 5′ end of the binding site of primer F1c. LoopB (LB) primers are positioned between the 5′ end of the binding site of primer B2 and the 3′ end of the binding site of primer B1c. In case loop primers are designed manually, make sure both loop primers have a melting temperature (T_m) of ±60 °C. In cases in which the available sequences are suitable only for the design of one possible loop primer (LF or LB), it should be used in any event to accelerate the LAMP reaction speed to a certain extent relative to the reaction with no loop primers.

12. Make sure to wear latex or nitrile gloves when preparing and handling LAMP master mixes. Do not reuse gloves after removal. Dispose DNA-contaminated gloves in an area remote from the area in which LAMP master mixes are prepared or handled. Use a dedicated lab coat in the master mix preparation area of your lab.

13. For the preparation of pure genomic DNA from fungal mycelia, we used the protocol of [55] with slight modifications: Mycelia were cultured in 1 mL of 2% malt extract broth in sterile 2 mL reaction tubes for 2–3 days at ambient temperature (AT) on a rotary shaker in horizontal position. Grown cultures were washed twice with 500 μL sterile tap water after centrifugation at 12,400× *g* for 5 min at AT. A 300 μL extraction buffer (200 mM Tris-HCl, 250 mM sodium chloride, 25 mM EDTA, 0.5% (w/v) SDS) as well as 0.5 g of sterile sea sand and 0.1 g of sterile glass beads (Ø 0.5 mm) were added before treatment in a FastPrep homogenizer (45 s; 24*2; 5.5 m/s, MP Biomedicals Germany GmbH, Eschwege, Germany). A 150 μL 3 M sodium acetate (pH 5.2) was added and the tube was placed at −20 °C for 10 min before centrifugation at 12,470× *g*, 5 min, AT. The supernatant was transferred to a fresh tube and mixed with an equal volume of isopropanol before placing the tubes at AT for 5 min for DNA precipitation. Following centrifugation (12,470× *g*, 5 min, AT), the DNA pellet was washed with 500 μL ice-cold 70% ethanol. Caps were inverted on a paper towel to run off the ethanol. Dry pellets were covered with 50 μL sterile ultrapure water and treated for 10 min at 50 °C until the DNA has dissolved. The extracted DNA was stored at 4 °C. The DNA concentration was measured spectrophotometrically using a Nanodrop 1000 instrument.
14. If no gradient thermal cycler is available, run three LAMP reactions as prepared according to Subheading 3.4 for 90 min at a constant block temperature of 65 °C. If all reactions turn into a bright pink color, repeat with fresh reactions at 66 °C or higher until no color change is visible after 90 min. If no color change occurs at 65 °C, repeat with fresh reactions at 64 °C or lower until color change occurs after 90 min. Use the highest temperature under which color change occurs as optimum temperature for further experiments.
15. Since the *Bst* polymerase used in LAMP reactions does not provide a proofreading activity, sequence errors can occur in the resulting product DNA with a rate higher than in PCR products. Therefore, it can be anticipated that base changes will occur in the LAMP product with a relatively high frequence. Sequencing of the smallest LAMP product will therefore result in a nucleotide sequence that will not necessarily show 100% identity with the sequence of the target gene. However, we assume that sequence identity over wide stretches of the product sequence verifies the identity of the LAMP product with the targeted gene sequence.

16. In cases where the sensitivity of a LAMP assay is very high or the initial concentration of the reference template DNA is very high, it can be necessary to test dilutions higher than 10^{-7}. Moreover, it is important to know the initial DNA concentration of the reference template in order to calculate the concentration of template DNA present in each of the LAMP reactions as a measure for assay sensitivity. We tend to use DNA mass per reaction (mass/rxn) as measure of assay sensitivity because this value is independent from the template concentration in the initial DNA extract.

17. Note that assay sensitivities in group-specific LAMP assays can vary for the different members of the group of target fungi. Also, the quality of the template DNA used for sensitivity testing may have an influence. We have tested assay sensitivity only with DNA of *A. parasiticus* CBS 126.62 as reference to provide a rough estimate for assay sensitivity.

18. The taxonomy and nomenclature of fungi are generally quite complicated and this is especially the case in the genus *Aspergillus* and related taxa. It is therefore strongly recommended to involve a specialist in fungal taxonomy and/or mycotoxins in order to pick a reasonable set of species for specificity testing. Besides testing closely related species, we recommend to test also more distantly related species that can potentially occur on natural sample materials, e.g., *Penicillium* spp., *Alternaria* spp., *Fusarium* spp., or species that are known plant pathogens occurring on the material to be analyzed with the assay. We have tested the current LAMP assay with a total of 128 different fungal species, representing 28 different genera. Among these species, those marked with a "+" in Table 2 were the only ones showing a positive result in the aflD-specific LAMP assay.

19. *Aspergillus oryzae* and *A. sojae* are domesticated species that currently are viewed as non-aflatoxigenic clones of *A. flavus* and *A. parasiticus*, respectively. Both domesticated forms are widely used in the production of Asian food such as soy sauce. Both are assumed to be safe for food production due to their inability to produce aflatoxins. However, analyses for aflalatoxin and CPA production have revealed that some *A. oryzae* strains were able to produce aflatoxins and cyclopiazonic acid (CPA, [56]). In *A. sojae*, Chang [57] found failure of interaction between the aflatoxin pathway-specific regulatory gene transcripton factor *aflR* and the coactivator protein *aflJ* as the reason for the general inability of that species to produce aflatoxins. Both species can be assumed to have an *aflD* (=*nor1*) gene so that they will be detected by the current LAMP assay. However, both types are not frequently found on natural materials and will therefore cause only few false-positive results.

20. Direct testing of pure culture material for the presence or absence of the *aflD* gene can be a useful tool for the screening of vast numbers of cultures isolated from naturally infected materials that might be prone to aflatoxin contamination. Instead of culturing strains under specific conditions and analyzing each culture for the presence or absence of aflatoxins, direct testing with the *aflD*-specific LAMP assay may help to select only those strains with an aflatoxigenic potential for further analysis. This strategy can save considerable amounts of time, money, and lab resources.
21. The aim of the experiment was to demonstrate that the developed LAMP assay can be used as a simple and rapid tool for the screening of a wider number of samples with varying characteristics and processing stages that might occur in a food quality lab or during a food safety surveillance campaign. The same protocol for LAMP template preparation was applied for samples of whole seeds of rice, maize, peanuts, and Brazil nuts; whole raisins and figs; ground hazelnuts and almonds that had previously been made into a slurry for mycotoxin analysis; and dry powdered paprika and ginger. In order to analyze whether or not a positive/negative LAMP result correlated with the presence/absence of aflatoxins in corresponding samples, most of the tested materials had previously undergone mycotoxin and/or mycological analysis by sample providers.
22. Table 3 shows that among 33 analyzed samples the number of samples in which the absence of a LAMP signal corresponded with the absence of aflatoxins was 21 (=95.5% correct negatives), whereas only one sample with negative LAMP signal was tested positive for aflatoxins (=4.5% false negatives). Within 11 samples with a positive LAMP result, six were also positive for aflatoxins (54.5% correct positives), whereas five were negative for the toxin (45.5% false positives). However, it should be noted here that even though aflatoxin production may not have occurred in all samples with a positive LAMP result, the detection of the responsible fungi will enable prediction of the potential risk of an aflatoxin contamination. Based on our data, the results translate into an 80.8% specificity of the assay for the detection of aflatoxin-contaminated samples in rice.

References

1. Cullen JM, Newberne PM (2013) Acute hepatotoxicity. In: Eaton DL, Groopman JD (eds) The toxicology of aflatoxins: human health, veterinary, and agricultural significance. Academic Press, San Diego
2. Saha Turna N, Wu F (2022) Estimation of tolerable daily intake (tdi) for immunological effects of aflatoxin. Risk Anal 42:431–438
3. Eaton DL, Gallagher EP (1994) Mechanisms of aflatoxin carcinogenesis. Ann Rev Pharmacol Toxicol 34:135–172

4. Milićević DR, Škrinjar M, Baltić T (2010) Real and perceived risks for mycotoxin contamination in foods and feeds: challenges for food safety control. Toxins 2:572–592
5. Egmond HP, Schothorst RC, Jonker MA (2007) Regulations relating to mycotoxins in food. Anal Bioanal Chem 1:147–157
6. Jallow A, Xie H, Tang X, Qi Z, Li P (2021) Worldwide aflatoxin contamination of agricultural products and foods: from occurrence to control. Comp Rev Food Sci Food Saf 20: 2332–2381
7. Ellis W, Smith JP, Simpson BK, Oldham JH, Scott PM (1991) Aflatoxins in food: occurrence, biosynthesis, effects on organisms, detection, and methods of control. Crit Rev Food Sci Nutr 30:403–439
8. Cotty PJ, Jaime-Garcia R (2007) Influences of climate on aflatoxin producing fungi and aflatoxin contamination. Int J Food Microbiol 119:109–115
9. Varga J, Frisvad JC, Samson R (2011) Two new aflatoxin producing species, and an overview of *Aspergillus* section *Flavi*. Stud Mycol 69:57–80
10. Frisvad JC, Hubka V, Ezekiel CN, Hong SB, Nováková A, Chen AJ, Arzanlou M, Larsen , TO, Sklenář F, Mahakarnchanakul W (2019) Taxonomy of *Aspergillus* section *Flavi* and their production of aflatoxins, ochratoxins and other mycotoxins. Stud Mycol 93:1–63
11. Pickova D, Ostry V, Malir F (2021) A recent overview of producers and important dietary sources of aflatoxins. Toxins 13:186
12. Silva JJ, Fungaro MHP, Wang X, Larsen TO, Frisvad JC, Taniwaki MH, Iamanaka BT (2022) Deep genotypic species delimitation of *Aspergillus* section *Flavi* isolated from Brazilian Foodstuffs and the description of *Aspergillus annui* sp. nov. and *Aspergillus saccharicola* sp. nov. J Fungi 8:1279
13. Bennett J, Papa KE, Ingram DS, Williams PA (1988) The aflatoxigenic *Aspergillus* spp. Adv Plant Pathol: Gen Plant Path Fungi 6:263–280
14. Olsen M, Johnsson P, Möller T, Paladino R, Lindblad M (2008) *Aspergillus nomius*, an important aflatoxin producer in Brazil nuts? World Myc J 1:123–126
15. Massi FP, Vieira Carneiro ML, Sartori D, Silva Penha RE, de Freitas MC, Ferreira JM, Iamanaka BT, Taniwaki MH, Frisvad JC, Fungaro Pelegrinelli MH (2014) Brazil nuts are subject to infection with B and G aflatoxin-producing fungus, *Aspergillus pseudonomius*. IntJ Food Microbiol 186:14–21
16. Copetti MV, Iamanaka BT, Pereira JL, Fungaro MH, Taniwaki MH (2011) Aflatoxigenic fungi and aflatoxin in cocoa. Int J Food Microbiol 148:141–144
17. Frisvad JC et al (2019) Taxonomy of *Aspergillus* section *Flavi* and their production of aflatoxins, ochratoxins and other mycotoxins. Stud Mycol 93:1–63
18. Bartoli A, Maggi O (1978) Four new species of *Aspergillus* from Ivory Coast soil. Trans Brit Mycol Soc 71:383–394
19. Frisvad JC, Skouboe P, Samson RA (2005) Taxonomic comparison of three different groups of aflatoxin producers and a new efficient producer of aflatoxin B 1, sterigmatocystin and 3-O-methylsterigmatocystin, *Aspergillus rambellii* sp. nov. Syst Appl Microbiol 28:442–453
20. Frisvad JC, Samson R, Smedsgaard J (2004) *Emericella astellata*, a new producer of aflatoxin B1, B2 and sterigmatocystin. Lett Appl Microbiol 38:440–445
21. Frisvad JC, Samson RA (2004) *Emericella venezuelensis*, a new species with stellate ascospores producing sterigmatocystin and aflatoxin B 1. Syst Appl Microbiol 27:672–680
22. Zalar P, Frisvad JC, Gunde-Cimerman N, Varga J, Samson RA (2008) Four new species of *Emericella* from the Mediterranean region of Europe. Mycologia 100:779–795
23. Chen A, Frisvad JC, Sun BD, Varga J, Kocsubé S, Dijksterhuis J, Kim DH, Hong SB, Houbraken J, Samson RA (2016) *Aspergillus* section *Nidulantes* (formerly *Emericella*): polyphasic taxonomy, chemistry and biology. Stud Mycol 84:1–118
24. Pitt J, Taniwaki MH, Cole M (2013) Mycotoxin production in major crops as influenced by growing, harvesting, storage and processing, with emphasis on the achievement of food safety objectives. Food Contr 32:205–215
25. Tai B, Chang J, Liu Y, Xing F (2020) Recent progress of the effect of environmental factors on *Aspergillus flavus* growth and aflatoxins production on foods. Food Qual Saf 4:21–28
26. Klich M (2002) Identification of common *Aspergillus* species. Centraalbureau voor Schimmelcultures, Utrecht
27. Sweeney MJ, Pàmies P, Dobson AD (2000) The use of reverse transcription-Polymerase Chain Reaction (RT-PCR) for monitoring aflatoxin production in *Aspergillus parasiticus* 439. Int J Food Microbiol 56:97–103
28. Rodríguez A, Rodríguez M, Luque MI, Martín A, Córdoba JJ (2012) Real-time PCR assays for detection and quantification of aflatoxin-producing molds in foods. Food Microbiol 31:89–99

29. Bu R, Sathiapalan RK, Ibrahim MM, Al-Mohsen I, Almodavar E, Gutierrez MI, Bhatia K (2005) Monochrome LightCycler PCR assay for detection and quantification of five common species of *Candida* and *Aspergillus*. J Med Microbiol 54:243–248
30. Cruz P, Buttner MP (2008) Development and evaluation of a real-time quantitative PCR assay for *Aspergillus flavus*. Mycologia 100:683–690
31. Sardiñas N, Vázquez C, Gil-Serna J, González-Jaén MT, Patiño B (2011) Specific detection and quantification of *Aspergillus flavus* and *Aspergillus parasiticus* in wheat flour by SYBR® Green quantitative PCR. Int J F Microbiol 145:121–125
32. Shapira R et al (1996) Detection of aflatoxigenic molds in grains by PCR. Appl Environ Microbiol 62(9):3270–3273
33. Geisen R (1996) Multiplex Polymerase Chain Reaction for the detection of potential aflatoxin and sterigmatocystin producing fungi. Syst Appl Microbiol 19(3):388–392
34. Färber P, Geisen R, Holzapfel W (1997) Detection of aflatoxinogenic fungi in figs by a PCR reaction. Int J Food Microbiol 36(2–3): 215–220
35. Manonmani H et al (2005) Detection of aflatoxigenic fungi in selected food commodities by PCR. Process Biochem 40(8):2859–2864
36. Mayer Z et al (2003) Quantification of the copy number of *nor*-1, a gene of the aflatoxin biosynthetic pathway by real-time PCR, and its correlation to the CFU of *Aspergillus flavus* in foods. Int J Food Microbiol 82(2):143–151
37. Rahimi P, Sharifnabi B, Bahar M (2008) Detection of aflatoxin in *Aspergillus* species isolated from pistachio in Iran. J Phytopathol 156(1): 15–20
38. Mideros SX et al (2009) *Aspergillus flavus* biomass in maize estimated by quantitative real-time Polymerase Chain Reaction is strongly correlated with aflatoxin concentration. Plant Dis 93(11):1163–1170
39. Passone MA et al (2010) Detection and quantification of *Aspergillus* section *Flavi* spp. in stored peanuts by real-time PCR of *nor*-1 gene, and effects of storage conditions on aflatoxin production. Int J Food Microbiol 138(3):276–281
40. Ahmad MM et al (2014) Detection of *Aspergillus flavus* and *Aspergillus parasiticus* from aflatoxin-contaminated peanuts and their differentiation using PCR-RFLP. Ann Microbiol 64(4):1597–1605
41. Mangal M et al (2016) Validation of PCR based detection system for aflatoxin producing molds. Indian J Exp Biol 54:472–476
42. Rossen L et al (1992) Inhibition of PCR by components of food samples, microbial diagnostic assays and DNA-extraction solutions. Int J Food Microbiol 17(1):37–45
43. Schrader C et al (2012) PCR inhibitors–occurrence, properties and removal. J Appl Microbiol 113(5):1014–1026
44. Notomi T et al (2000) Loop-mediated isothermal amplification of DNA. Nucleic Acids Res 28(12):e63–e63
45. Niessen L et al (2013) The application of Loop-mediated isothermal amplification (LAMP) in food testing for bacterial pathogens and fungal contaminants. Food Microbiol 36(2):191–206
46. Parida M et al (2008) Loop mediated isothermal amplification (LAMP): a new generation of innovative gene amplification technique; perspectives in clinical diagnosis of infectious diseases. Rev Med Virol 18(6):407–421
47. Niessen L (2015) Current state and future perspectives of Loop-mediated isothermal amplification (LAMP)-based diagnosis of filamentous fungi and yeasts. Appl Microbiol Biotechnol 99(2):553–574
48. Tasrip N et al (2019) Loop mediated isothermal amplification; a review on its application and strategy in animal species authentication of meat based food products. Int Food Res J 26:1–10
49. Li JJ et al (2016) Loop-mediated isothermal amplification (LAMP): emergence as an alternative technology for herbal medicine identification. Front Plant Sci 7:1956
50. Kaneko H et al (2007) Tolerance of Loop-mediated isothermal amplification to a culture medium and biological substances. J Biochem Biophys Methods 70(3):499–501
51. Goto M et al (2009) Short technical reports – colorimetric detection of Loop-mediated isothermal amplification reaction by using hydroxy naphthol blue. BioTechniques 46: 167–172
52. Tanner NA, Zhang Y, Evans TC Jr (2015) Visual detection of isothermal nucleic acid amplification using pH-sensitive dyes. BioTechniques 58(2):59–68
53. Yu J et al (2004) Clustered pathway genes in aflatoxin biosynthesis. Appl Environ Microbiol 70:1253–1262
54. Georgianna DR, Payne GA (2009) Genetic regulation of aflatoxin biosynthesis: from gene to genome. Fungal Genet Biol 46:113–125
55. Cenis J (1992) Rapid extraction of fungal DNA for PCR amplification. Nucleic Acids Res 20(9):2380

56. Kim NY et al (2014) An evaluation of aflatoxin and cyclopiazonic acid production in *Aspergillus oryzae*. J Food Prot 77(6):1010–1016
57. Chang P-K (2004) Lack of interaction between AFLR and AFLJ contributes to nonaflatoxigenicity of Aspergillus sojae. J Biotechnol 107(3):245–253
58. Niessen L et al (2018) LAMP-based group specific detection of aflatoxin producers within *Aspergillus* section *Flavi* in food raw materials, spices, and dried fruit using neutral red for visible-light signal detection. Int J Food Microbiol 266:241–250

Chapter 16

Fecal Contamination Assessment Through Loop-Mediated Isothermal Amplification of *Bacteroidales* in Food Samples

Jiangshan Wang and Mohit S. Verma

Abstract

Fecal contamination of food samples originating from both human and animal sources poses a significant public health risk. Traditional laboratory techniques require specialized equipment and skilled workers, including a time-consuming enrichment process that makes them unsuitable for field testing. Here we describe a rapid risk-assessment assay for fecal contamination by targeting *Bacteroidales* using loop-mediated isothermal amplification (LAMP). The assay allows for naked-eye observation of reactions, detecting as few as eight copies of *Bacteroidales* per cm^2 on the surface of food samples within 50 min.

Key words Isothermal nucleic acid amplification, *Bacteroidales*, Color change, Food safety

1 Introduction

Contamination of food by fecal matter, originating from both human and animal sources, poses a public health risk associated with the potential for foodborne illnesses [1, 2]. Current methods for identifying fecal contamination are deemed insufficient or inefficient. One prevalent method is via visual inspection, where workers scrutinize the surface of the product, such as fresh produce, for signs of fecal contamination during the harvesting process. Items found to be contaminated are discarded, while those with no evident contamination are collected. Furthermore, in slaughterhouses, a stringent visual criterion mandates that all surfaces of processed meat and poultry carcasses be free of obvious fecal contamination before entering the carcass chiller. However, it is difficult for workers to thoroughly inspect the surfaces of each product since some contamination is invisible or barely visible and can be missed. Another conventional procedure for microbiological analyses is a culture-based method. Spot-check samples are delivered to a microbiology lab and cultured with diverse mediums for detecting enteric pathogens or quantifying fecal indicator bacteria (FIB).

Alejandro Garrido-Maestu and Alexandre Lamas (eds.), *Foodborne Pathogens: Loop-mediated Isothermal Amplification Protocols*, Methods and Protocols in Food Science, https://doi.org/10.1007/978-1-0716-4526-0_16,

Although the test is cost-effective (excluding shipping costs) and straightforward for trained personnel, culture-based methods require a minimum of 1–3 days to yield results, delaying discoveries and preventing early warnings and prompt implementation of contamination control or mitigation steps [3–5]. Culture-based approaches necessitate access to a microbiology lab and have limitations in detecting the viable but non-culturable (VBNC) state [6, 7].

To quickly determine microbial contamination in-field without requiring access to a laboratory, professional staff, or expensive equipment, here we demonstrate a rapid risk-assessment assay for fecal contamination that targets *Bacteroidales* using loop-mediated isothermal amplification (LAMP) [8]. Due to the unique characteristics of LAMP *Bst* DNA polymerase, LAMP enables simpler detection of microorganisms in environmental samples, requiring only a single temperature (in the range of 60–65 °C) and exhibiting resistance to common polymerase chain reaction (PCR) inhibitors found in unpurified environment samples, enabling direct measurements [9]. Incorporating a colorimetric dye (e.g., EBT, phenol red) in LAMP assays enables color changes that are visible to the naked eye [10]. With minimal training, nonspecialist users can use these tests in the field with basic equipment to determine site-specific fecal contamination. This assay can assist stakeholders in their decision-making process regarding the microbial safety of food samples.

2 Materials

2.1 Sample Compatibility

The liquid-based LAMP assay is designed to accommodate samples in liquid format. To enhance the concentration of the target DNA prior to conducting the assay, we suggest utilizing a swab/cloth to thoroughly sample the designated area, followed by resuspending the collected material in nucleic-free water [11–13]. The resuspension can be directly used for the LAMP assay. For samples with a liquid nature, such as wash solutions, it is recommended to employ membrane filtration to concentrate DNA on the membrane [14]. Subsequently, resuspend the membrane in a smaller volume of nucleic acid-free water. In this chapter, we will use a customized collection flag and cellulose swab to demonstrate the assay process within fresh produce fields. It is worth noting that alternative methods can be employed to accommodate other scenarios and to ensure optimal performance for other applications.

The collection flag is made with a bamboo skewer (29.8 cm) and a strip of transparent film (Apollo Plain Paper Copier Transparency Film) [11]. The transparent film is pre-cut into strips measuring 7.62 × 21.59 cm (3 × 8.5 inches). Each strip is stapled at the edge to form a loop. A bamboo skewer was inserted through the loop to make a collection flag. Each collection flag is swabbed using a wet polyester-tipped swab (263,000, BD BBL, USA). Swabs from other suppliers should work as well.

2.2 LAMP Preparation

This study employed a commercialized LAMP master mix, although equivalent reagents from other suppliers should work as well.

1. Before you start: Apply RNase AWAY to pipettes and workbench surfaces. Wipe everything down completely. Residues will interfere with the reaction.
2. Thaw WarmStart® Colorimetric LAMP 2X Master Mix (M1800; New England Biolabs, USA) at room temperature. Salt may appear at the bottom of the tube; therefore vortex briefly or invert tubes multiple times to ensure proper mixing before use.
3. For each reaction, prepare the reaction mix as described below.

Components	Volume	Unit	Final concentration	Unit
WarmStart Colorimetric LAMP 2× Master Mix	12.5	μL	1	X
LAMP Primer Mix (10×)[a]	2.5	μL	1	X
Nuclease-free water	9	μL		

[a]We recommend making stocks of the LAMP primers at a workable concentration for ease of setup. We suggest making a 10× Primer Mix containing all six LAMP primers. 10× Primer mix: 16 μM FIP/BIP, 2 μM F3/B3, 4 μM Loop F/B

3 Methods

3.1 Deploying Collection Flags and Sample Processing

1. Collection flags are placed in the area of interest 7 days in advance. The fields are labeled with row and column numbers, and flags are placed at the intersection of each row and column. The flags can be deployed at varying densities depending on the size of the sampling area and the level of resolution desired. We used a consistent distance of 6 m between each row and column, resulting in approximately 100 sampling sites per acre of the field.
2. The collection flags are encoded with a unique identifier and the location associated with the flag's identifier is recorded.
3. After 7 days, all collection flags are collected and stored in individual Ziploc resealable storage bags (B07NQVYCG3, Amazon, USA).

4. Swab both sides of each collection flag using a wet polyester-tipped swab (263000, BD BBL, USA), and resuspend the swab in 200 μL of nuclease-free water.

5. The resuspension is ready for the LAMP assay. The samples can also be kept at −20 °C until the experiment.

3.2 LAMP Reaction

1. Pipette 24 μL per reaction into desired reaction vessels and add 1 μL of sample (add equivalent volume of nuclease-free water or sample storage buffer for non-template reactions).

2. Mix by pipetting. Check that reaction solutions have a bright pink color, which indicates the initial high pH required for a successful colorimetric LAMP reaction (*see* **Note 1**).

3. Seal reaction vessels and incubate at 65 °C for 45 min. The heat source can be any basic heating device that is capable of reaching 65 °C (e.g., incubator, water bath) (*see* **Notes 2** and **3**) [11, 13, 15, 16].

4. Remove tubes or vessels from incubation and visually inspect the results. Negative reactions should retain a pink color, while positive reactions should exhibit a color change to yellow. If the color change is not distinct, such as an orange color, return reactions to 65 °C for an additional 10 min. It is also possible to check reactions earlier. High-concentration reactions may display a complete color change in 10–15 min. The color can be observable immediately upon removal from the incubation temperature but can be enhanced by allowing the reaction to cool to room temperature (*see* **Note 4**).

4 Notes

1. When adding the sample to the reaction mixture, the solution may change from a bright pink to a red-violet or orange-yellow color. This color shift signals a substantial impact on the pH of the added sample to the reaction matrix. To address this, we recommend diluting the sample with nuclease-free water until its addition no longer alters the initial color.

2. During the incubation step, negative reactions may change color concurrently or shortly after sample-containing reactions. This issue can be caused by either primer dimerization/nonspecific amplification or carryover contaminants of previous LAMP reactions. To prevent carryover contamination, we recommend the following: (1) use deoxynucleotide triphosphates (dUTPs) and uracil-DNA glycosylase (UDG) in all LAMP reactions; (2) keep separate working locations for LAMP mixture preparation and sample addition; and (3) aliquot reagent stocks and replace them if contamination is suspected. Over-incubating the reaction

would result in nonspecific amplification. We recommend a maximum incubation duration of 60 min.

3. LAMP is less sensitive to inhibitors present in the field samples compared to qPCR and thus can amplify the targets better with non-extracted samples. However, if the sample is excessively complex (e.g., blood, urine, or saliva), we recommend diluting the sample or heat-inactivating it at 90 °C for 5 min.
4. The assay could be integrated into a microfluidic paper-based analytical device. By preloading amplification reagents onto a paper-based biosensor and incorporating a smart design, a multistep amplification process can be completed on one chip [15, 17]. This approach enhances user-friendliness, especially for individuals without specialized expertise.

Acknowledgments

This work was funded by the Center for Produce Safety (CPS Award Number: 2021CPS12), the California Department of Food and Agriculture (CDFA Agreement No. 20-0001-054-SF), and the US Department of Agriculture's (USDA) Agricultural Marketing Service (USDA Cooperative Agreement No. USDA-AMS-TM-SCBGP-G-20-0003). Any opinions, findings, conclusions, or recommendations expressed in this publication or audiovisual are those of the author(s) and do not necessarily reflect the views of The Center for Produce Safety, the California Department of Food and Agriculture, or the US Department of Agriculture's (USDA) Agricultural Marketing Service. The project entitled "Field evaluation of microfluidic paper-based analytical devices for microbial source tracking" was funded in whole or in part through a subrecipient grant awarded The Center for Produce Safety through the California Department of Food and Agriculture 2020 Specialty Crop Block Grant Program and the US Department of Agriculture's (USDA) Agricultural Marketing Service.

References

1. Alegbeleye OO, Singleton I et al (2018) Sources and contamination routes of microbial pathogens to fresh produce during field cultivation: a review. Food Microbiol 73:177–208. https://doi.org/10.1016/j.fm.2018.01.003
2. Hoelzer K, Moreno Switt AI, Wiedmann M, Boor KJ (2018) Emerging needs and opportunities in foodborne disease detection and prevention: from tools to people. Food Microbiol 75:65–71. https://doi.org/10.1016/j.fm.2017.07.006
3. Foddai ACG, Grant IR (2020) Methods for detection of viable foodborne pathogens: current state-of-art and future prospects. Appl Microbiol Biotechnol 104:4281–4288. https://doi.org/10.1007/s00253-020-10542-x
4. Holcomb DA, Stewart JR (2020) Microbial indicators of fecal pollution: recent progress and challenges in assessing water quality. Curr Environ Health Rep 7:311–324. https://doi.org/10.1007/s40572-020-00278-1
5. Noble RT, Weisberg SB (2005) A review of technologies for rapid detection of bacteria in recreational waters. J Water Health 3:381–392. https://doi.org/10.2166/wh.2005.051

6. Li Y, Huang T-Y, Ye C, Chen L, Liang Y, Wang K, Liu J (2020) Formation and control of the viable but non-culturable state of foodborne pathogen *Escherichia coli* O157:H7. Front Microbiol 11
7. Zhao X, Zhong J, Wei C, Lin C-W, Ding T (2017) Current perspectives on viable but non-culturable state in foodborne pathogens. Front Microbiol 8
8. Notomi T, Okayama H, Masubuchi H, Yonekawa T, Watanabe K, Amino N, Hase T (2000) Loop-mediated isothermal amplification of DNA. Nucleic Acids Res 28: 63e. https://doi.org/10.1093/nar/28.12.e63
9. Francois P, Tangomo M, Hibbs J, Bonetti E-J, Boehme CC, Notomi T, Perkins MD, Schrenzel J (2011) Robustness of a Loop-mediated isothermal amplification reaction for diagnostic applications. FEMS Immunol Med Microbiol 62:41–48. https://doi.org/10.1111/j.1574-695X.2011.00785.x
10. Tanner NA, Zhang Y, Evans TC (2015) Visual detection of isothermal nucleic acid amplification using pH-sensitive dyes. BioTechniques 58:59–68. https://doi.org/10.2144/000114253
11. Wang J, Ranjbaran M, Ault A, Verma MS (2023) A Loop-mediated isothermal amplification assay to detect *Bacteroidales* and assess risk of fecal contamination. Food Microbiol 110: 104173. https://doi.org/10.1016/j.fm.2022.104173
12. Wang J, Ranjbaran M, Verma MS (2024) *Bacteroidales* as a fecal contamination indicator in fresh produce industry: a baseline measurement. J Environ Manag 351:119641. https://doi.org/10.1016/j.jenvman.2023.119641
13. Wang J, Kaur S, Kayabasi A, Ranjbaran M, Rath I, Benschikovski I, Raut B, Ra K, Rafiq N, Verma MS (2024) A portable, easy-to-use paper-based biosensor for rapid in-field detection of fecal contamination on fresh produce farms. Biosens Bioelectron 259:116374. https://doi.org/10.1016/j.bios.2024.116374
14. Jiang YS, Riedel TE, Popoola JA, Morrow BR, Cai S, Ellington AD, Bhadra S (2018) Portable platform for rapid in-field identification of human fecal pollution in water. Water Res 131:186–195. https://doi.org/10.1016/j.watres.2017.12.023
15. Davidson JL, Wang J, Maruthamuthu MK, Dextre A, Pascual-Garrigos A, Mohan S, Putikam SVS, Osman FOI, McChesney D, Seville J, Verma MS (2021) A paper-based colorimetric molecular test for SARS-CoV-2 in saliva. Biosens Bioelectron X 9:100076. https://doi.org/10.1016/j.biosx.2021.100076
16. Pascual-Garrigos A, Maruthamuthu MK, Ault A, Davidson JL, Rudakov G, Pillai D, Koziol J, Schoonmaker JP, Johnson T, Verma MS (2021) On-farm colorimetric detection of *Pasteurella multocida*, *Mannheimia haemolytica*, and *Histophilus somni* in crude bovine nasal samples. Vet Res 52:126. https://doi.org/10.1186/s13567-021-00997-9
17. Wang J, Dextre A, Pascual-Garrigos A, Davidson JL, Maruthamuthu MK, McChesney D, Seville J, Verma MS (2021) Fabrication of a paper-based colorimetric molecular test for SARS-CoV-2. MethodsX 8:101586. https://doi.org/10.1016/j.mex.2021.101586

Part V

Novel Applications

Chapter 17

Implementation of Fluorescent Primers in Loop-Mediated Isothermal Multiplex Assays

Patrick Hardinge

Abstract

Loop-mediated amplification (LAMP) is an example of an isothermal nucleic acid amplification technology. LAMP uses template-specific primers to provide sufficient amplified DNA for detection and quantification. LAMP has numerous benefits over PCR including single temperature reaction conditions enabling the use of low-complexity equipment, high sensitivity, robustness to certain inhibitors, and rapid amplification. However, the specificity of LAMP can be problematic despite the number of primers involved, and false-positive results can occur from nonspecific primer interactions. Detection and quantification methods that record increasing total DNA are therefore inferior to methods that track specifically the LAMP reaction such as the quenched fluorescent primer technique.

In this chapter, assay design considerations for LAMP quenched fluorescent primers are described and a multiplexed method for testing with multiple fluorophores is shown.

Key words Quenched fluorescent primers, Loop-mediated amplification (LAMP), Genotyping, DNA, Isothermal nucleic acid amplification technology (iNAAT), Förster resonance energy transfer (FRET), Multiplex

1 Introduction

In this chapter, the application of quenched fluorescent primers is introduced for the isothermal nucleic acid amplification technology, loop-mediated amplification (LAMP) [1]. The aim of this method is to provide high specificity, rapid nucleic acid amplification, and detection and quantification of foodborne pathogen DNA.

The mechanism of loop-mediated isothermal amplification (LAMP) is more complex than PCR due to the number of primers involved. Displacement primers (F3 and B3) are required to strand invade the double-stranded DNA template to start the amplification. LAMP primers (forward inner primer FIP and backward inner primer BIP) form the hairpin loops that are required for the LAMP stages of cycling and elongation (*see* Fig. 1). Further primers can be

Alejandro Garrido-Maestu and Alexandre Lamas (eds.), *Foodborne Pathogens: Loop-mediated Isothermal Amplification Protocols*, Methods and Protocols in Food Science, https://doi.org/10.1007/978-1-0716-4526-0_17,

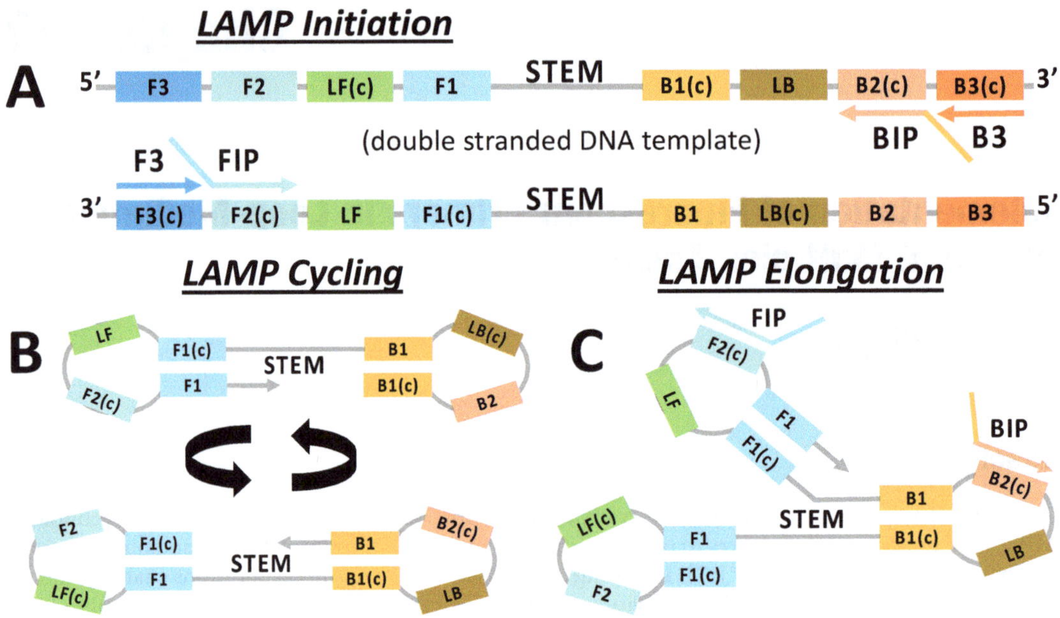

Fig. 1 LAMP reaction key stages: (**a**) The initiation of the LAMP amplification by displacement and LAMP primers that strand invade the double-stranded DNA target sequence. A polymerase with displacement activity creates the nascent strands with loops in "dumbbell" formations that are required for the cycling stage (**b**). Further priming with LAMP primers from the "dumbbells" creates highly looped structures in the elongation stage (**c**). Displacement primer sequences are termed F3 and B3 with reference to their forward or backward priming activity. Loop primer sequences are LF and LB, and LAMP primers (FIP and BIP) are composed of two sequences F1, F2 and B1, B2). Complementary sequences are marked with (**c**)

utilized to accelerate the reaction such as loop primers (LF and LB) [2], which anneal to the newly formed loops and STEM primers [3], which anneal to the stem region between F1 and B1 (*see* Fig. 1). LAMP amplifies at a constant temperature typically between 60 and 65 °C, without the requirement to thermocycle. LAMP can also tolerate certain PCR inhibitors [4]. During the elongation stage of LAMP, complex cauliflower-like loop structures are formed, which are essential to the quenched fluorescent primer method.

A constant temperature typically between 60 and 65 °C depending on the assay is required by LAMP and this can be achieved with a heat block. That are numerous detection methods for real-time and end-point LAMP with changes in fluorescence being the most common. Detection requires excitation of fluorescent dyes, such as SYTO9 or SYBR Green, or fluorophores, for example, FAM (6-carboxylfluoroscein), JOE (6-carboxy-4′,5′-dichloro-2′,7′-dimethyoxyfluoroscein) and TAMRA (6-carboxytetramethylrhodomine), and recording the emission fluorescence with appropriate filters. Quantitative PCR devices can be used for this such as a Qiagen RotorGene with multiple excitation and emission channels.

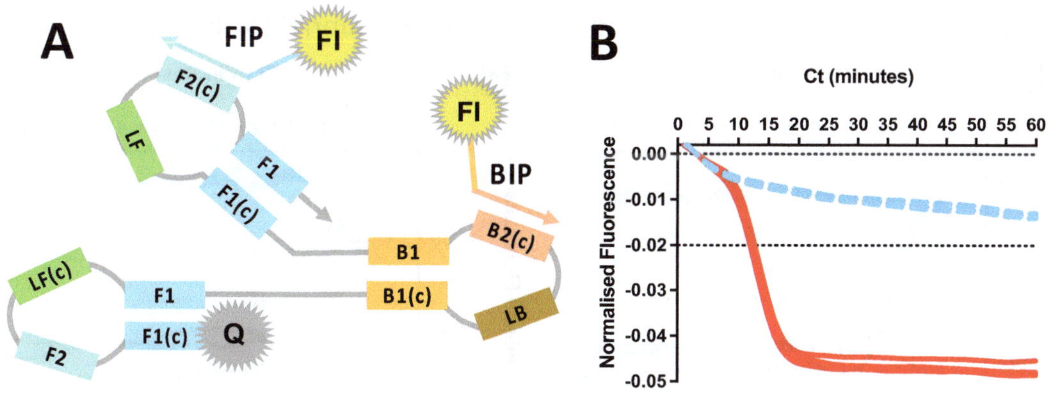

Fig. 2 Quenched fluorescent primer LAMP. (**a**) Representation of a LAMP amplicon in the early stages of elongation to show fluorescent LAMP primers FIP and BIP annealing to complementary sequences for further nascent DNA polymerisation. The fluorophores on these primers are unquenched (Fl). The incorporated fluorophore (Q) is quenched by FRET due to its proximity to energy receiving nucleotides. An example trace is shown in (**b**) with positive results in red and no template controls (NTCs) in blue. The fluorescence of the positive sample is quenched (reduced) during LAMP amplification

Quenched fluorescent primer LAMP [5] utilizes certain fluorophores attached to the 5′ end of LAMP primers (FIP and BIP) that are quenched by LAMP amplification (*see* Fig. 2). The quenching of the fluorophore is possibly by proximity to guanine nucleotides [6] as the amplicons elongate into cauliflower-like structures. The proximity enables Förster resonance energy transfer (FRET) to occur and the fluorescent signal to reduce in intensity. A number of technologies have been devised that use FRET with LAMP amplification including modified primer methods such as duplex LAMP [7] and fluorescently labeled loop primers [8, 9] and fluorescent probe-based methods such as LAMP-OSD [10], DARQ [11], and QUASR [12].

The quenched fluorescent primer assay duration is dependent on template concentration and the reaction kinetics of the assay; lower concentrations of DNA template provide longer cycle threshold (Ct) times. Therefore, the technique can be used to quantify the initial DNA concentration of the template.

LAMP amplification can be affected by nonspecific primer interactions giving false-positive results. Highly specific methods that detect true LAMP amplification from the utilization of FIP and BIP primers are therefore preferable. Using SYTO9 dye to detect increasing DNA concentration and JOE-labeled LAMP primers to monitor LAMP, it was possible to separate false-positive results from LAMP results [5]. This could be beneficial to LAMP primer design. Another benefit is the potential of quenched fluorescent LAMP for the detection of multiple targets in the same tube by multiplexing.

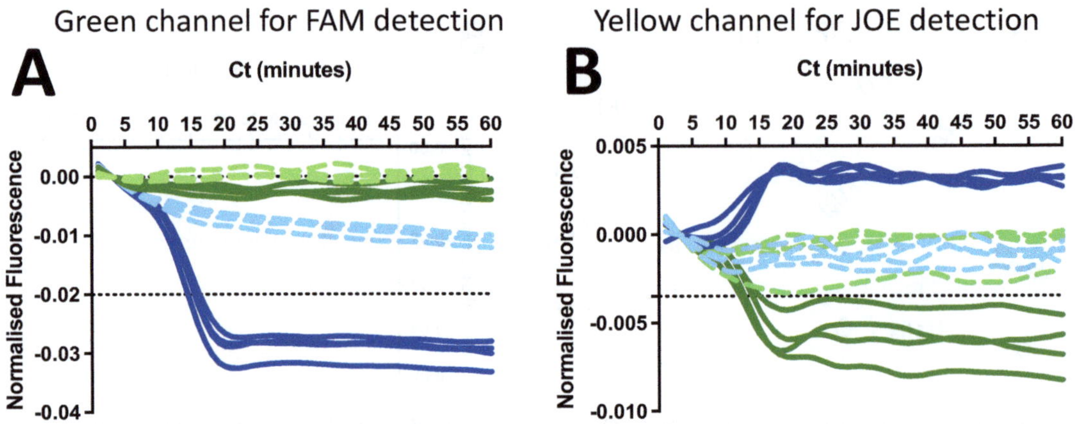

Fig. 3 Example of multiplexed quenched fluorescent primer LAMP. The combined DNA templates for two pathogens are specifically amplified with a FAM-labeled LAMP assay for pathogen 1 (FAM-labeled BIP primer) and a JOE-labeled LAMP assay for pathogen 2 (JOE labelled FIP primer). (**a**) FAM-labeled LAMP amplication of pathogen 1 is detected by the green channel of the RotorGene thermocycler (blue) with NTCs in light blue. (**b**) In the same tubes, the yellow channel detects the JOE-labeled LAMP amplification of pathogen 2 (green) with NTCs in light green

Using fluorophores that can be detected at distinctly different wavelengths associated with different sets of LAMP primers (*see* Fig. 3) enables multiplexing. Previously this has been demonstrated [5] with FAM excitation 495 nm/emission 520 nm and TAMRA 544 nm excitation/576 nm emission in a duplex system. In the example in Fig. 3, the duplex reaction utilized FAM with JOE, which has excitation at 520 nm and emission at 545 nm.

2 Materials

LAMP amplification rapidly produces high concentrations of amplified template DNA. LAMP is also highly sensitive to DNA template contamination (sample and amplified); therefore, stringent anti-contamination procedures are required (*see* **Note 1**). Unless otherwise stated, all reagents are prepared at room temperature (RT).

2.1 Requirements for Quenched Fluorescent Primer LAMP

1. Design of the quenched primers: The FIP or BIP primer of the existing LAMP primer set can be modified by adding a fluorophore to the 5′ end at synthesis. No spacer is required, and quenching has still been observed when the fluorophore is attached to a guanine nucleotide. The fluorophores FAM, JOE, and TAMRA have been used for quenching, but it is possible that other fluorophores may also function with the appropriate channel selection on the thermocycler for detection.

2. Modifications for the LAMP quenched primer assay: There is no requirement for a fluorescent dye such as SYTO9 or SYBR Green to detect the DNA synthesized during LAMP amplification. However, using SYTO9 with a JOE-labeled LAMP assay on a multichannel thermocycler can be useful to assess "false-positive" results from nonspecific primer interactions that will be indicated by positive results by SYTO9 detection only [5]. Carrier DNA often used to enhance LAMP reactions can be omitted from quenched fluorescent primer LAMP.

2.2 Reagents for Quenched Flourescent Primer LAMP

1. Isothermal Buffer (*see* **Note 2**): Supplied at 10 × concentration for direct usage. Store at −20 °C.
2. DTT (1,4-dithiothreitol): Weigh out 1.54 g of DTT in a falcon tube and add 10 mL of molecular grade water. Thoroughly mix and prepare 100 μL labeled single-use aliquots of 1 M DTT. Store at −20 °C.
3. PVP (polyvinylpyrrolidone): Weigh out 0.4 g of PVP in a falcon tube and add 10 mL of molecular grade water. Thoroughly mix and prepare 100 μL labeled aliquots of 40 mg/mL PVP. Store at −20 °C.
4. KCl (potassium chloride): Weigh out 0.895 g of KCl in a falcon tube and add 10 mL of molecular grade water. Thoroughly mix and prepare 100 μL labeled aliquots of 1.2 M KCl. Store at −20 °C.
5. dNTPs (deoxyribonucleotide triphosphates) (*see* **Note 3**): Add 10 μL dATP, dCTP, dGTP, and dTTP to 160 μL of molecular grade water. Mix and prepare 65 μL labeled aliquots of 5 mM dNTPs. Store at −20 °C.
6. Betaine: Supplied at 5 M concentration and used directly to give a final concentration of 0.8 M (*see* **Note 4**). Stored in fridge at 4 °C.
7. *Bst* polymerase (*see* **Note 5**): Supplied at 8000 U/μL and used directly to give a final concentration of 0.32 U/μL. Stored at −20 °C.
8. LAMP primers (*see* **Note 6**): Prepared fresh at the time of the LAMP assay from defrosted stocks of each primer. To make 100 μL 10 × primer mix, add 2 μL each displacement primer (F3 and B3), 4 μL each loop primer (LF and LB), and 8 μL each LAMP primer (FIP and BIP) to 72 μL of molecular grade water.

2.3 Consumables and Specialist Equipment

1. Non-stick 1.5 mL tubes (*see* **Note 7**).
2. Maximum recovery pipette tips (*see* **Note 7**).
3. qPCR tubes or 96-well microtitre plate (*see* **Note 8**).
4. Vortex.

5. Microcentrifuge.
6. Temperature-controlled device with fluorescence excitation source and optical filters such as a RotorGene qPCR thermocycler (*see* **Note 9**).

2.4 Consumables for Genomic Extraction

1. 10 mL of lysis buffer (containing 2 mL of 0.5 M EDTA, 2.5 mL of 25% SDS, 200 μL of 1 M Tris pH 8.0, and 5.3 mL of molecular grade water).
2. Proteinase K: 100 μg/mL.
3. RNase A: 20 μg/mL.
4. Freshly prepared 70% ethanol solution.
5. Phenol/chloroform/isoamyl alcohol.
6. 3 M sodium acetate
7. 100% ethanol
8. Samples: In the examples in Fig. 3, two parasites of cattle were used: rumen fluke and a liver fluke (*Calicophoron daubneyi* and *Fasciola hepatica*). DNA extracted from these organisms were tested with the quenched fluorescent primer LAMP method [13].
9. Mortar and pestle.
10. Vortex.
11. Microcentrifuge.
12. Non-stick 1.5-ml tubes (*see* **Note 7**).
13. Maximum recovery pipette tips (*see* **Note 7**).
14. Heatblock or incubator at 37 °C.
15. Chemical safety hood for handling of phenol/chloroform/isoamyl alcohol.

3 Methods

The preparation of the LAMP assay must be in an area where contamination from DNA template is controlled (*see* **Note 1**). All procedures can take place at RT if warm-start polymerase is used, unless otherwise specified. If another type of polymerase is used, the preparation of the LAMP mix should be on ice. This LAMP method has previously been described in published protocols [14, 15].

3.1 Preparation of LAMP Primers

1. To prepare 150 μL of the primer mix for 75 LAMP reactions of 20 μL, thaw the 100 μM LAMP primers (FIP and BIP), loop (LF and LB), and displacement primers (F3 and B3), and briefly vortex and pulse spin.

2. Add 3 μL of each displacement primer, 6 μL of each loop primer, and 12 μL of each LAMP primer to a non-stick 1.5 mL tube using fresh pipette tips. Return the primers to storage at −20 °C.
3. Add 108 μL of molecular grade water, briefly vortex, and pulse spin. Label the primer mix.
4. Keep the primer mix on ice until use.

3.2 Preparation of DNA Samples

1. Pulverize and lyse the cellular material, for example, the adult stage of *Fasciola hepatica* and *Calicophoron duabneyi*, with 500 μL of the lysis buffer, and incubate overnight at 37 °C.
2. Add 20 μL of Proteinase K and 3 μL RNase A and incubate for 2 h at 37 °C.
3. In a safety hood, add an equal volume of phenol/chloroform/isoamyl alcohol (500 μL) and vortex at full speed for 20 s.
4. Centrifuge at 12000 rpm for 5 min, remove, and discard the aqueous layer.
5. Add 50 μL 3 M sodium acetate.
6. Divide and add two to three times the volume of 100% ethanol, vortex, and transfer to a − 20 °C freezer for a minimum of 1 h.
7. Centrifuge immediately at 14000 rpm for 15 min, and remove the supernatant carefully to keep the pellet.
8. Wash the pellet with 1 ml of 70% ethanol. Centrifuge at 14000 rpm for 5 min and remove the supernatant. Repeat the washing process five times.
9. Air-dry the pellet before the addition of 100 μL of molecular grade water, rehydrate, and store overnight at 4 °C.

3.3 Preparation of LAMP Mix

1. For 72 reactions of 20 μL reaction volumes (if using a Rotor-Gene qPCR device), prepare 1125 μL of LAMP mix for 15 μL loads with 5 μL samples according to the following protocol. Ensure the required LAMP assay components are thawed, briefly vortex, and pulse spin.
2. To a non-stick 1.5 mL tube, add 272.5 μL of molecular grade water and 150 μL of the 10 × Isothermal Buffer slowly.
3. Add 15 μL DTT to give a final concentration of 10 mM.
4. Slowly draw up and slowly add 15 μL PVP for a final concentration of 400 μg/mL.
5. Add 75 μL KCl to give a final concentration of 60 mM.
6. Add 90 μL of the combined dNTPS for a final concentration of 300 μM with respect to each dNTP.
7. Add 240 μL Betaine solution to give a final concentration of 0.8 M.

8. Add 60 μL of warm-start *Bst* polymerase to give a final concentration of 0.32 U/μL. Slowly draw up and add the polymerase due to the viscosity.
9. Add 150 μL of the LAMP primer mix and gently pipette up and down to ensure homogeneity of the reactants (*see* **Note 10**).

3.4 Quenched Fluorescent Primer LAMP Assay

1. To 15 μL of the LAMP reaction mix, add 5 μL of sample or NTC as required.
2. Transfer the tubes to the heat block with fluorescence detection and start the reaction and data collection.
3. Set the assay parameters depending on assay optimization: temperature (between 60 and 65 °C), acquisition interval, and length of assay (between 30 and 60 min).

4 Notes

1. A single copy of DNA template can be amplified using the LAMP method [16]; therefore DNA template and amplified DNA pose a high risk of contamination that requires mitigation. The risk of contamination can be reduced by preparing single-use aliquots of the key LAMP reagents in advance in batches that can be traced. The LAMP assay preparation time can be reduced with this measure. A dedicated clean area, for reagent preparation and assay setup that is free from amplified DNA and other nucleic acid templates, is an essential requirement. A 70% ethanol solution from spray bottles should be used to wipe surfaces and a UV crosslinking is useful to reduce contamination on pipettes and plasticware. The risk of contamination can be reduced by using a Class II Biological Safety Cabinet to prepare reagents, LAMP primers, and LAMP assays. Dedicated equipment, reagents, and plasticware should be used for these procedures. To monitor contamination within assays, negative template controls (NTCs) should be included in assay designs.
2. The 10 × Isothermal Buffer can be prepared in-house with 20 mM Tris-HCl, 10 mM $(NH_4)_2SO_4$, 50 mM KCl, 2 mM $MgSO_4$, and 0.1% Tween 20 at pH 8.8 (at 25 °C) or purchased from New England Biolabs (NEB). Before use, fully defrost the 10 × Isothermal Buffer and mix thoroughly with a vortex to ensure any precipitate returns to solution.
3. The deoxyribonucleotide triphosphates (dNTPs) are diluted from 100 mM stock (available from Invitrogen) by adding 10 μL of each dNTP to 160 μL molecular grade water for 5 mM with respect to each dNTP. For assays with a large total volume, each dNTP can be used directly for a final concentration of each dNTP of 300 μM.

4. A 5 M Betaine solution can be purchased and used directly. The benefit of 0.8 M betaine in PCR and LAMP assays is the subject of debate [17, 18]. It is used to to enhance amplification by decreasing the stability of GC base pairs, but it may not be required for LAMP assays.
5. *Bst* polymerase can be purchased from NEB as a warm-start enzyme with an aptamer inhibiting the polymerase activity below 45 °C. This feature allows LAMP assays to be set up at RT. This warm-start *Bst* polymerase can be used for LAMP, but alternative versions, enzymes, and suppliers are available.
6. The synthesized oligonucleotide LAMP primers are stored at a stock concentration of 100 μM at −20 °C. The 10 × LAMP primer mix contains 8 μM of both the forward inner primer (FIP) and backward inner primer (BIP), 4 μM of forward loop (LF) and backward loop (LB), and 2 μM of forward displacement primer (F3) and backward displacement primer (B3). Twice the concentration of the LAMP primers (FIP and BIP) or twice the concentration of all the primers can also be used and may enhance the LAMP reaction.
7. During LAMP preparation there is a risk that the DNA template may bind to the plasticware that could reduce the availability of the template for amplification. Non-stick 1.5-mL tubes and maximum recovery pipette tips are recommended throughout LAMP preparation to reduce this risk especially if working with low copy number template.
8. Ensure that the appropriate 0.1 mL tubes and caps are used if using a RotorGene thermocycler for a LAMP assay. Ensure the tip of the pipette is angled and deep into the tube when loading the reaction mix to prevent loss. If using 96-well microtiter plates, ensure that they are suitable for the heat block wells and design.
9. For a RotorGene thermocycler, the parameters should be set to acquire to channels corresponding to the fluorophore of the quenched primers (e.g., the yellow channel for the fluorophore JOE and green channel for FAM). Define the assay temperature (between 60 and 65 °C), length of assay (between 30 and 60 min depending on the LAMP assay and concentration of DNA template), and acquisition time (between 30 and 60 s). Alternative devices can be used that allow these parameters to be set.
10. Using a vortex to mix the LAMP reaction components increases the risk of bubble and microbubble formation in the tube. Alternatively a pipette can be used to provide thorough mixing by slowly drawing up and down the solution repeatedly. Bubbles and microbubbles could interfere with the LAMP reaction and fluorescence detection.

Acknowledgments

The research into quenched fluorescent primers in LAMP assays was funded by BBSRC Stand Alone LINK grant BB/L022346/1, a BBSRC Industrial CASE award BB/G017336/1, and the European Regional Development Fund with Cardiff University 80762-CU-085 (East).

References

1. Notomi T, Okayama H, Masubuchi H, Yonekawa T, Watanabe K, Amino N et al (2000) Loop-mediated isothermal amplification of DNA. Nucleic Acids Res 28:E63. https://doi.org/10.1093/nar/28.12.e63
2. Nagamine K, Hase T, Notomi T (2002) Accelerated reaction by Loop-mediated isothermal amplification using loop primers. Mol Cell Probes 16:223–229. https://doi.org/10.1006/mcpr.2002.0415
3. Gandelman O, Jackson R, Kiddle G, Tisi L (2011) Loop-mediated amplification accelerated by stem primers. Int J Mol Sci 12:9108–9124. https://doi.org/10.3390/ijms12129108
4. Kiddle G, Hardinge P, Buttigieg N, Gandelman O, Pereira C, McElgunn CJ et al (2012) GMO detection using a bioluminescent real time reporter (BART) of loop mediated isothermal amplification (LAMP) suitable for field use. BMC Biotechnol 12:15. https://doi.org/10.1186/1472-6750-12-15
5. Hardinge P, Murray JAH (2019) Reduced false positives and improved reporting of Loop-mediated isothermal amplification using quenched fluorescent primers. Sci Rep 9:7400. https://doi.org/10.1038/s41598-019-43817-z
6. Torimura M, Kurata S, Yamada K, Yokomaku T, Kamagata Y, Kanagawa T et al (2001) Fluorescence-quenching phenomenon by photoinduced electron transfer between a fluorescent dye and a nucleotide base. Anal Sci 17(1):155–160. https://doi.org/10.2116/analsci.17.155
7. Kouguchi Y, Fujiwara T, Teramoto M, Kuramoto M (2010) Homogenous, real-time duplex Loop-mediated isothermal amplification using a single fluorophore-labeled primer and an intercalator dye: its application to the simultaneous detection of Shiga toxin genes 1 and 2 in Shiga toxigenic *Escherichia coli* isolates. Mol Cell Probes 24(4):190–195. https://doi.org/10.1016/j.mcp.2010.03.001
8. Zerilli F, Bonanno C, Shehi E, Amicarelli G, Adlerstein D, Makrigiorgos GM (2010) Methylation-specific Loop-mediated isothermal amplification for detecting hypermethylated DNA in simplex and multiplex formats. Clin Chem 56(8):1287–1296. https://doi.org/10.1373/clinchem.2010.143545
9. Spinelli O, Rambaldi A, Rigo F, Zanghì P, D'Agostini E, Amicarelli G et al (2014) Simple, rapid and accurate molecular diagnosis of acute promyelocytic leukemia by Loop-mediated amplification technology. Oncoscience 2(1):50–58. https://doi.org/10.18632/oncoscience.114
10. Jiang YS, Bhadra S, Li B, Wu YR, Milligan JN, Ellington AD (2015) Robust strand exchange reactions for the sequence-specific, real-time detection of nucleic acid amplicons. Anal Chem 87(6):3314–3320. https://doi.org/10.1021/ac504387c
11. Tanner NA, Zhang Y, Evans TC Jr (2012) Simultaneous multiple target detection in real-time Loop-mediated isothermal amplification. BioTechniques 53(2):81–89. https://doi.org/10.2144/0000113902
12. Ball CS, Light YK, Koh CY, Wheeler SS, Coffey LL, Meagher RJ (2016) Quenching of unincorporated amplification signal reporters in reverse-transcription Loop-mediated isothermal amplification enabling bright, single-step, closed-tube, and multiplexed detection of RNA viruses. Anal Chem 88(7):3562–3568. https://doi.org/10.1021/acs.analchem.5b04054
13. Hardinge P (2025) Novel genome assembly of *Calicophoron daubneyi* for LAMP molecular diagnostic fieldable assay. BioRxiv preprint submission
14. Hardinge P (2022) Optimized Loop-mediated isothermal amplification (LAMP) allows single copy detection using bioluminescent assay in real time (BART). Methods Mol Biol 2524:107–117. https://doi.org/10.1007/978-1-0716-2453-1_8

15. Hardinge P (2023) Molecular beacons – Loop-mediated amplification (MB-LAMP). Methods Mol Biol 2638:289–299. https://doi.org/10.1007/978-1-0716-3024-2_20
16. Hardinge P, Kiddle G, Tisi L, Murray JAH (2028) Optimised LAMP allows single copy detection of 35Sp and NOSt in transgenic maize using Bioluminescent Assay in Real Time (BART). Sci Rep 8(1):17590. https://doi.org/10.1038/s41598-018-36207-4
17. Ma C, Wang Y, Zhang P, Shi C (2017) Accelerated isothermal nucleic acid amplification in betaine-free reaction. Anal Biochem 1(530): 1–4. https://doi.org/10.1016/j.ab.2017.04.017
18. Henke W, Herdel K, Jung K, Schnorr D, Loening SA (1997) Betaine improves the PCR amplification of GC-rich DNA sequences. Nucleic Acids Res 25(19):3957–3958. https://doi.org/10.1093/nar/25.19.3957

Chapter 18

Same-Day Detection of Viable *Salmonella* Enteritidis by Phage Amplification Coupled with Loop-Mediated Isothermal Amplification

Alexandre Lamas and Alejandro Garrido-Maestu

Abstract

In recent years, isothermal techniques such as loop-mediated isothermal amplification (LAMP) have appeared as alternatives to qPCR. Due to its amplification at constant temperature, LAMP reactions can be performed in equipment as simple as a thermoblock. However, one of the main limitations of nucleic acid-based techniques is their incapacity to differentiate between live and death cells. To overcome this limitation, some researchers are evaluating the detection of bacteriophages (phages) DNA instead of the detection of bacterial nucleic acids. Phages are used in the sample enrichment step where they reproduce in the presence of live cells. Due to narrow spectrum activity, one phage can infect one bacterial species or even serotype, which allows to perform a selective enrichment. This chapter describes a protocol for the detection of *Salmonella* Enteritidis in chicken meat in less than 8 h. This method combines sample enrichment in peptone water to which phage PVP-SE2 is added with phage DNA detection by LAMP. In addition, a colorimetric variant of this technique is described that allows this technique to be applied at points of interest.

Key words *Salmonella*, Chicken meat, LAMP, Phage, Point of care

1 Introduction

Salmonella spp. is a major foodborne pathogen worldwide [1]. Poultry meat, eggs, and derived products are one of the main sources of *Salmonella* in the food production chain [2]. One of the main serotypes transmitted from these food products is *S.* Enteritidis, which in 2022 was responsible for 67.3% of human salmonellosis cases in the European Union [3]. *Salmonella* spp. detection methods continue to be mainly based on culture-dependent methods, such as ISO 6579:2017. However, these protocols present various limitations such as the long times to obtain the final result and the necessary labor [4]. In recent years, molecular techniques, such as qPCR, have emerged as an interesting alternative to these

Alejandro Garrido-Maestu and Alexandre Lamas (eds.), *Foodborne Pathogens: Loop-mediated Isothermal Amplification Protocols*, Methods and Protocols in Food Science, https://doi.org/10.1007/978-1-0716-4526-0_18,

culture-based methods. Although qPCR can be considered a gold standard within nucleic acid amplification techniques, it has certain limitations, such as the need for specialized equipment that requires an important initial investment [5]. This can limit its application in small-capacity laboratories and even in the food industry itself. To overcome these limitations, molecular techniques based on isothermal amplification such as loop-mediated isothermal amplification (LAMP) are a promising alternative [6, 7]. These techniques are characterized by using a constant amplification temperature and therefore the reactions can be carried out in a simple thermoblock or even in a water bath. Additionally, this technique is compatible with different chemicals that allow for naked-eye detection, making the technique even more attractive for point-of-interest analyses [8].

However, one of the main limitations of nucleic acid-based methods assays is their inability to differentiate between viable and nonviable microorganisms. To overcome this limitation, various alternatives have been sought, such as the use of DNA intercalating dyes, which block the amplification of DNA from dead bacteria such as propidium monoazide [9] or the detection of messenger RNA instead of DNA [10]. A more recent, promising alternative is the combination of molecular methods with bacteriophages (phages), viruses that specifically infect and replicate in bacterial cells. Phages are characterized by a narrow infection spectrum, being only able to infect a bacterial species, a specific serotype, or even a strain [11]. This feature allows the development of methods for pathogen-specific detection, based on the amplification of phage genomic DNA that reproduces in the presence of the host rather than target the pathogen DNA directly [12]. The present chapter describes a method based in phage amplification coupled with LAMP for same-day detection of viable cells of *S.* Enteritidis in raw poultry meat.

2 Materials

2.1 Culture Media

1. Buffered peptone water (BPW): 10 g peptone, 5 g NaCl, Na_2HPO_4 $12H_2O$, KH_2PO_4, 1000 mL distilled water. Mix well by stirring. Autoclave at 121 °C for 15 min.
2. Tryptic soy broth: 17 g digest pancreatic of casein 3 g papaic digest of soya, 2.5 g D(+)-glucose, 5 g sodium chloride, 2.5 g di-potassium hydrogen phosphate, 15 g agar, 1000 mL distilled water. Mix well by stirring. Autoclave at 121 °C for 15 min.
3. Semi-solid tryptic soy broth: 17 g digest pancreatic of casein 3 g papaic digest of soya, 2.5 g D(+)-glucose, 5 g sodium chloride, 2.5 g di-potassium hydrogen phosphate, 7 g agar, 1000 mL distilled water. Mix well by stirring. Autoclave at 121 °C for 15 min.

2.2 Chemicals and Materials

1. SM buffer (100 mM NaCl, 50 mM Tris-HCl, 8 mM $MgSO_4 \cdot 7H_2O$, pH 7.5).
2. Fast Master Mix (ISO-004, OptiGene, Horsham, UK).
3. SYBR Green I Solution (Invitrogen™, ThermoFisher Scientific, Waltham, MA, USA) (*see* **Note 1**).
4. ROX, passive reference dye (Invitrogen™, ThermoFisher Scientific, Carlsbad, CA, USA).
5. Molecular biology grade water (nuclease-free).
6. LAMP primers (Table 1).
7. 96-well PCR-compatible microplates.
8. Adhesive optical quality film for 96-well microplates, qPCR compatible.
9. 0.2 mL eight-tube PCR strips.
10. Laboratory film. Parafilm® M was used in this work.
11. Sterile homogenizer bags with filter.

2.3 Equipment

1. Scale.
2. Laboratory homogenizer as Stomacher® 400 Circulator or SMASHER® Sample Blender.
3. Laboratory incubator suitable for 37 °C.
4. Refrigerated centrifuge with a rotor suitable for 1.5–2 mL microtubes and centrifugation capacity up to 16,000× *g*.
5. Heating block with agitation capacity capable to reach 99 °C. In the described workflow, a Thermomixer 5436 (Eppendorf AG, Germany) was used.
6. Real-time PCR thermocycler with filter for SYBR® Green or equivalent dye with similar emission wavelength. In the described methodology, a QuantStudio™ 5 (Applied Biosystems, ThermoFisher Scientific, Waltham, MA, USA) was used.

Table 1
LAMP primers

Phage gene	Primer name	Sequence (5′- > 3′)
Endolysin	FIP-Ph_end-Salm	GTCTACCGCGGCTACAGCCTTA-GAGGGCCAGAAGATTACCGA
	BIP-Ph_end-Salm	GCGCATCCGTCGCTCAATCA-CCAGCGTTATACACCAGGTC
	LF-Ph_end-Salm	CCCTGTTCAGTAGCAAAAGACCCTG
	LB-Ph_end-Salm	GTCACAGTTCGATGCGATGTGC
	F3-Ph_end-Salm	AATTGGCTACGGCCACTAC
	B3-Ph_end-Salm	GAAGCCGCAATCACACCG

7. Conventional PCR thermocycler. In this work Veriti Thermal Cycler (Applied Biosystems, ThermoFisher Scientific, CA, USA) was used.

3 Methods

3.1 Phage Selection and Primer Design

1. Select a phage with lytic activity against the target bacteria. In the present work, PVP-SE2 phage, which host is *S.* Enteritidis, was selected [13, 14] (*see* **Note 2**). *S.* Enteritidis S1400 from the private collection of the University of Bristol was used as host strain to develop the method.
2. Select a region of the phage genome for primer design. In this case, the annotated genome sequence of PVP-SE2 phage was downloaded from NCBI database with the accession number MF431252.1. Endolysin gene was selected for primer design (*see* **Note 3**). Endolysin gene in Fasta format was uploaded to Primer Explorer V5 (https://primerexplorer.jp/lampv5e/index.html) and primers were designed.

3.2 Microbiological Analysis and DNA Extraction

1. Preheat BPW at 37 °C in a laboratory incubator or in a water bath (*see* **Note 4**).
2. Weigh 25 g of chicken meat in a stomacher bag with filter.
3. Add 225 mL of preheated BPW to the sample, and homogenize for at least 1 min.
4. Incubate the sample at 37 °C for 3 h.
5. Add the phage to a final concentration of 10^4 PFU/ mL, and incubate the sample at 37 °C for another 4 h (*see* **Note 5**). Phage has to be diluted to an appropriate concentration from stock using SM buffer.
6. For DNA extraction, transfer 1 mL of sample to a 1.5–2 mL microtube, and heat it at 99 °C under constant agitation (1400 rpm) for 5 min (*see* **Note 6**).
7. Centrifuge the sample at 16,000× *g* for 1 min at 4 °C to eliminate food and cellular debris, and transfer the supernatant with the DNA to a new 1.5 mL microtube. The samples will be placed at 4 °C, or −20 °C, for short- or long-term storage, respectively (Fig. 1).

3.3 Real-Time LAMP

1. Prepare the LAMP reaction mixture according to Table 2, and load the samples in 96-well qPCR plates. Seal the plates with adhesive optical quality film (*see* **Note** 7).
2. Load the plates in a qPCR instrument. The thermal profile is described in Table 3.

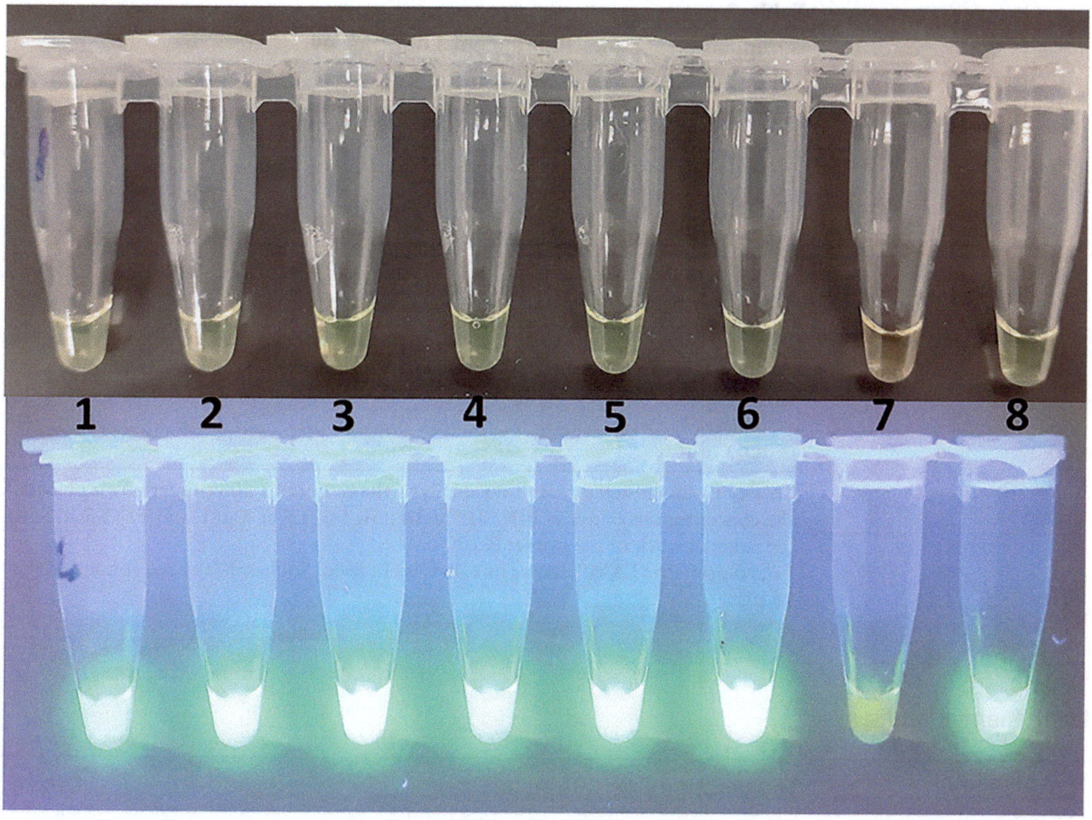

Fig. 1 Colorimetric LAMP observed with the naked eye (upper image) and under UV light (below image). From one to six positive samples, seven negative control, and eight positive control

3.4 Colorimetric LAMP

1. The reaction mixture is the same described in Table 3 but without ROX (*see* **Note 8**). Dispense the mix, and the sample, in 0.2 mL PCR tubes. With the lid open, cover the tubes with Parafilm® M leaving an opening of 1 mm on the hinge side of the tube. Then place in the lid 1 μL of 1000× SYBR Green I solution, and close the tube carefully.
2. Place the tube in a thermocycler at 66 °C for 30 min (*see* **Note 9**). After amplification, shake the tube to allow the mix of the sample with the SYBR Green I solution placed in the lid. Then, spin the tubes in a benchtop centrifuge, so the whole liquid moves to the bottom of the tube to better visualize the result.

3.5 Result Analysis and Interpretation

1. In the case of real-time LAMP, the first step is to determine the presence of amplification curves. If there are no curves, the sample is considered negative. If there are curves, the next step is to confirm the result by melt curve analysis to determine the melting temperature (Tm). For the endolysin target, the Tm obtained was 90.1 ± 0.31 °C. Therefore, positive samples will have an amplification curve and a melting peak around 90.1 °C. Samples with amplification curve but with a different melting temperature are considered negative.

Table 2
Composition for one fluorescent LAMP assay

Component	Volume/rxn (μL)	Final concentration (nM)
LAMP master mix	12	–
Primer Mix[a]	1	FIB/BIP: 800 LB/LF: 400 F3/B3: 200
ROX[b]	0.3	50
Template DNA	2	–
H_2O	4.7	–
Final volume	20	–

[a]To avoid adding each primer individually to the reaction, a 20× primer mix is prepared. The concentration of mix is FIB/BIP at 16 μM, LB/LF at 8 μM, and F3/B3 at 4 μM. The original stock of the primer is 100 μM

[b]For colorimetric LAMP master mix, Rox is not used. So, in this case the volume of H_2O is 5 μL

Table 3
LAMP thermal profile

LAMP step	Temperature (°C)	Time (s)	Cycles
Isothermal amplification	66	30	30
Melt curve analysis	95 85 95	1 20 Temperature increases 0.05 °C/s from 85 °C	1 1 1

For colorimetric LAMP, a single step at 66 °C for 30 min is used

2. In the case of colorimetric LAMP, in positive samples SYBR Green I binds to double-stranded amplicons and the reaction will develop a greenish color visible by naked-eye observation and emits fluorescence when observed under a UV light. In negative samples, reactions remain orange and there is no fluorescence under UV light.

4 Notes

1. Other chemical substances that bind to DNA can be used as SYTO 9, SYTO 82, and propidium iodide. The only requirement is that a positive reaction causes a color change that can be detected with the naked eye or even under ultraviolet light. It must be taken into consideration whether the concentration of

intercalating molecule required to produce a color change detectable by the naked eye can cause an inhibition of the amplification. If this is the case, the substance must be added at the end of the reaction.

2. Phage selection is one of the most important points to develop a successful phage amplification method coupled with LAMP. Phages are characterized by its specificity, normally infecting only one species, serotype, or even some strains [11]. Therefore, it is important to select a phage that demonstrated a wide range activity against the target microorganism but do not infect nontarget microorganisms. Also, the phage must be sequenced not only to design the primers but also to genomic characterize it. Some phages can be inserted in the genome of the host and therefore stop infecting the cell and producing progeny, remaining in a latent state [11]. The use of those has to be avoided selecting only the obligately lytic ones. The genomic analysis allows to determine whether the phage can be inserted into the host genome [15].
3. In the exclusivity assays, it is important to include phages, and bacterial strains, of the target and related microorganisms, especially when conserved genes such as endolysin and capsid genes are selected. As commented in previous note, phages can be inserted in the host genome as prophages. A large number of *Salmonella* strains have these prophages in their genome [16]. Therefore, the molecular method could be detecting genes of prophages located in the bacterial chromosome instead of lytic phage used.
4. In order to shorten incubation times, it is important that BPW is preheated to 37 °C. If it is added at room temperature, once the sample is placed in the incubator, it will take some time for the sample to reach the ideal temperature and will therefore affect the multiplication of the bacteria and therefore to the performance of the method [17]. In this protocol, BPW was used as is the pre-enrichment medium recommended in ISO 6579-1:2017. Phage infection and reproduction can be affected by media composition and some of them are quite demanding with respect to the composition of the media. Therefore, when developing such a method, it is important to ensure that the phage reproduces properly in the chosen enrichment medium.
5. This concentration is optimal for the phage used in this study. Maybe other phages require other concentrations based on their infection activity. To prepare a fresh phage stock, the following protocol can be followed. One hundred μL of bacterial host, in this case *S.* Enteritidis S1400, exponential phase culture is mixed with 10 μL of decimal dilutions of the stock

phage in SM buffer and 3 mL of semi-solid TSA medium (0.7% agar) and plated on TSA agar plates. The plates are incubated at 37 °C overnight. An isolated phage plate is taken and spread on the surface of a double layer agar prepared as described above. This allows the phage to reproduce. After incubating overnight at 37 °C, 3 mL of SM buffer is added to the plate and incubated at 4 °C and shaking, 90 rpm, overnight to resuspend the phage. After this period, the liquid is transferred to a 15 mL tube and chloroform is added to a final concentration of 10%. The tube is centrifuged at 9000× *g* at 4 °C for 10 min. The aqueous phase is collected and filtered with a 0.22 μM syringe filter. The filtrate corresponds to the phage stock. To determine the concentration of phage in this stock, decimal dilutions of the stock are prepared in SM buffer and double agar plates are prepared as described above. This phage stock will be stored at 4 °C.

6. This step can be performed without constant agitation, but its use contributes to a homogeneous heat distribution, especially at the beginning of the process. It is also important to note that due to the evaporation caused by this heating, the caps of the microtubes tend to open. To avoid this, parafilm can be used around the caps.
7. For this protocol, phage endolysin gene was selected. But other conserved genes as capsid genes can be also used.
8. For colorimetric LAMP was selected the method described by Sukphattanaudomchoke et al. [18] that add SYBR Green after LAMP reaction. Positive reactions change to green, while negative reactions have an orange color. Also, under UV light positive reactions have fluorescence. But colorimetric LAMP master mix available in the market can be also used. One option is the use of colorimetric master mix as WarmStart® Colorimetric LAMP 2X Master Mix (New England Biolabs, Ipswich, MA, USA) where positive reactions change the color of reaction from red to yellow due to a presence of a pH indicator in the master mix that detects the pH changes produced during LAMP amplification. However, in tests performed with this master mix on inoculated chicken samples, a color change from red to yellow was observed just after adding the DNA sample, possibly due to the presence of substances in the sample that caused this change. For this reason, its use was discarded. Other options are ISO-004-VIS Fast Isothermal Master Mix (visual detection) or IsoFast® Hot Start *Bst* Polymerase Colour.
9. In this workflow was used a thermocycler, but other much simpler equipment such as a heat block, like the one used for the DNA extraction step, could be used. In this case, however, the method would have to be reevaluated to determine whether the conditions could be the same as those used with the thermocycler.

References

1. Teklemariam AD, Al-Hindi RR, Albiheyri RS, Alharbi MG, Alghamdi MA, Filimban AAR et al (2023) Human salmonellosis: a continuous global threat in the farm-to-fork food safety continuum. Food Secur 12
2. Tan SJ, Nordin S, Esah EM, Mahror N (2022) *Salmonella* spp. in chicken: prevalence, antimicrobial resistance, and detection methods. Microbiol Res 13:691
3. The European Union One Health 2022 Zoonoses Report (2023). EFSA J 21(12)
4. Kabiraz MP, Majumdar PR, Mahmud MMC, Bhowmik S, Ali A (2023) Conventional and advanced detection techniques of foodborne pathogens: a comprehensive review. Heliyon 9(4):e15482
5. Ndraha N, Lin HY, Wang CY, Hsiao HI, Lin HJ (2023) Rapid detection methods for foodborne pathogens based on nucleic acid amplification: recent advances, remaining challenges, and possible opportunities. Food Chem Mol Sci 7:100183
6. Shang Y, Ye Q, Cai S, Wu Q, Pang R, Yang S et al (2021) Loop-mediated isothermal amplification (LAMP) for rapid detection of *Salmonella* in foods based on new molecular targets. LWT 142
7. Kreitlow A, Becker A, Schotte U, Malorny B, Plötz M, Abdulmawjood A (2021) Establishment and validation of a Loop-mediated isothermal amplification (LAMP) assay targeting the *ttrRSBCA* locus for rapid detection of *Salmonella* spp. in food. Food Control 126:126
8. Garrido-Maestu A, Prado M (2022) Naked-eye detection strategies coupled with isothermal nucleic acid amplification techniques for the detection of human pathogens. Compr Rev Food Sci Food Saf 21:1913
9. Zheng X, Wang Y, Gong WZ, Cai Q, Li J, Wu J (2022) Detection of *Escherichia coli*, *Pseudomonas aeruginosa*, *Salmonella* paratyphoid B, and *Shigella dysentery* in live *Bacillus licheniformis* products using propidium monoazide-real-time-quantitative polymerase chain reaction. Front Microbiol:13
10. Miao YJ, Xiong GT, Bai MY, Ge Y, Wu ZF (2018) Detection of live *Salmonella enterica* in fresh-cut vegetables by a TaqMan-based one-step reverse transcription real-time PCR. Lett Appl Microbiol 66(5):447
11. Lin DM, Koskella B, Lin HC (2017) Phage therapy: an alternative to antibiotics in the age of multi-drug resistance. World J Gastrointest Pharmacol Ther 8(3)
12. Ding Y, Huang C, Chen M, Wang J, Shao Y, Wang X (2023) Rapid and simultaneous detection of viable *S. aureus* and its penicillin susceptibility by phage amplification techniques in different food matrices. LWT:176
13. Santos SB, Fernandes E, Carvalho CM, Sillankorva S, Krylov VN, Pleteneva EA et al (2010) Selection and characterization of a multivalent *Salmonella* phage and its production in a nonpathogenic *Escherichia coli* strain. Appl Environ Microbiol 76(21):7338
14. Sillankorva S, Pleteneva E, Shaburova O, Santos S, Carvalho C, Azeredo J et al (2010) *Salmonella* Enteritidis bacteriophage candidates for phage therapy of poultry. J Appl Microbiol 108(4):1175
15. McNair K, Aziz RK, Pusch GD, Overbeek R, Dutilh BE, Edwards R (2018) Phage genome annotation using the RAST pipeline. Methods Mol Biol
16. Gao R, Naushad S, Moineau S, Levesque R, Goodridge L, Ogunremi D (2020) Comparative genomic analysis of 142 bacteriophages infecting *Salmonella enterica* subsp. *enterica*. BMC Genomics 21(1):374
17. Fachmann MSR, Löfström C, Hoorfar J, Hansen F, Christensen J, Mansdal S et al (2017) Detection of *Salmonella enterica* in meat in less than 5 hours by a low-cost and non-complex sample preparation method. Appl Environ Microbiol 83(5)
18. Sukphattanaudomchoke C, Siripattanapipong S, Thita T, Leelayoova S, Piyaraj P, Mungthin M et al (2020) Simplified closed tube Loop-mediated isothermal amplification (LAMP) assay for visual diagnosis of *Leishmania* infection. Acta Trop 212:105651

Chapter 19

Gold Nanoparticle-Based Loop-Mediated Isothermal Amplification for Rapid, Low-Cost, and Simple Detection of *Salmonella* in Coffee and Cream

Shiyao Wang, Eric Wai Him Lee, Jack Yiu Tat Yuen, Jackson Chit Shing Woo, Stryer Wong, Benjamin Wai Kit Chan, Shea Ping Yip, and Thomas Ming Hung Lee

Abstract

The critical public health risk caused by *Salmonella* necessitates the need for efficient detection methods. Conventional techniques based on plate counting and real-time polymerase chain reaction are time-consuming, costly, and labor-intensive, thus mainly performed in centralized laboratories. To meet the emerging need for decentralized/on-site testing, here we describe a rapid, low-cost, and simple method for *Salmonella* detection in coffee and cream by using gold nanoparticle-based loop-mediated isothermal amplification.

Key words *Salmonella* detection, Loop-mediated isothermal amplification, Gold nanoparticles, Coffee, Cream

1 Introduction

Salmonella infection imposes a considerable global health and economic burden [1]. Increased public awareness related to foodborne contamination has led to greater efforts to develop faster, cheaper, and simpler methods of *Salmonella* detection for an effective food safety control response. The gold standard detection method is bacteria culture based on plate counting, which typically involves pre-enrichment, selective enrichment, and isolation on selective agar media, followed by confirmation through additional biochemical or serological tests. The process is excessively time-consuming (5–7 days), costly, and labor-intensive, which is mainly performed in centralized microbiology laboratories [2]. In recent decades, a variety of molecular methods have been extensively utilized for *Salmonella* detection, offering significant advantages over

Alejandro Garrido-Maestu and Alexandre Lamas (eds.), *Foodborne Pathogens: Loop-mediated Isothermal Amplification Protocols*, Methods and Protocols in Food Science, https://doi.org/10.1007/978-1-0716-4526-0_19,

traditional culture techniques. Real-time polymerase chain reaction (real-time PCR or quantitative PCR "qPCR," in short PCR) is the current nucleic acid testing gold standard, which has been reported for *Salmonella* detection in food samples with a pre-enrichment step in buffered peptone water followed by a deoxyribonucleic acid (DNA) isolation/purification step [3]. The turnaround time is around 24 h, much faster than that for the culture method. However, the pre-PCR procedure is complicated, while PCR requires bulky and expensive instrument, Hence, PCR-based nucleic acid testing is still limited in centralized settings. Isothermal nucleic acid amplification techniques are promising candidates for decentralized testing thanks to their simple temperature control and fast amplification [4]. Among them, loop-mediated isothermal amplification (LAMP) has the advantages of (1) simple isothermal temperature control at ~65 °C, (2) fast amplification in 15–30 min, (3) high specificity by means of 4–6 primers that recognize 6–8 distinct regions of the target sequence, and (4) high tolerance to unpurified samples for direct amplification (without nucleic acid isolation/purification steps) [5]. LAMP result readout can be achieved by real-time turbidity measurement (production of magnesium pyrophosphate crystals in a positive LAMP sample; endpoint visual readout by the naked eye is also possible but only under special lighting condition [6]). Another LAMP result readout can be achieved by colorimetric method based on the use of pH-sensitive dye (e.g., phenol red, production of hydrogen ions in a positive LAMP sample, under unbuffered or weakly buffered condition [7]). It should be pointed out that the assay accuracy is affected by the turbidity and pH of the specimen for these two LAMP result readout methods. To address this issue, gold nanoparticle precipitation readout method for LAMP (Gold-LAMP) was developed [8]. The gold nanoparticles are surface-functionalized with thiolated poly(ethylene glycol) and 11-mercaptoundecanoic acid (PEG/MUA-AuNPs). For a negative Gold-LAMP sample (absence of a target nucleic acid sequence), the gold nanoparticles appear as a red dispersion. On the other hand, for a positive Gold-LAMP sample (presence of a target nucleic acid sequence), the gold nanoparticles turn to a red precipitate as a result of the coprecipitation with magnesium pyrophosphate crystals (Fig. 1). Herein, *Salmonella* detection in coffee (containing tannic acid) [9] and cream (containing fat) [10] is demonstrated, which is very challenging with PCR due to enzyme inhibition. Together with a simple and low-cost sample preparation method by filtration (Fig. 2; without the need for nucleic acid isolation/purification as well as complicated operation steps/instruments), *Salmonella* can be detected in enriched coffee and cream samples by Gold-LAMP assay (65 °C for 25 min) with a clear result readout by the naked eye (Fig. 3).

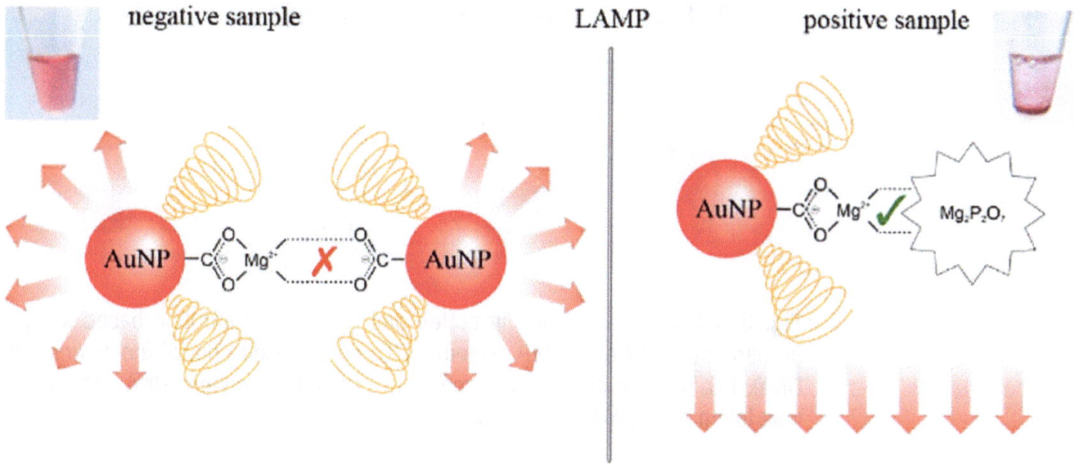

Fig. 1 Schematic illustration of the gold nanoparticle-based loop-mediated isothermal amplification (Gold-LAMP) assay. (Reproduced from Ref. [8] with permission from American Chemical Society, Copyright 2017)

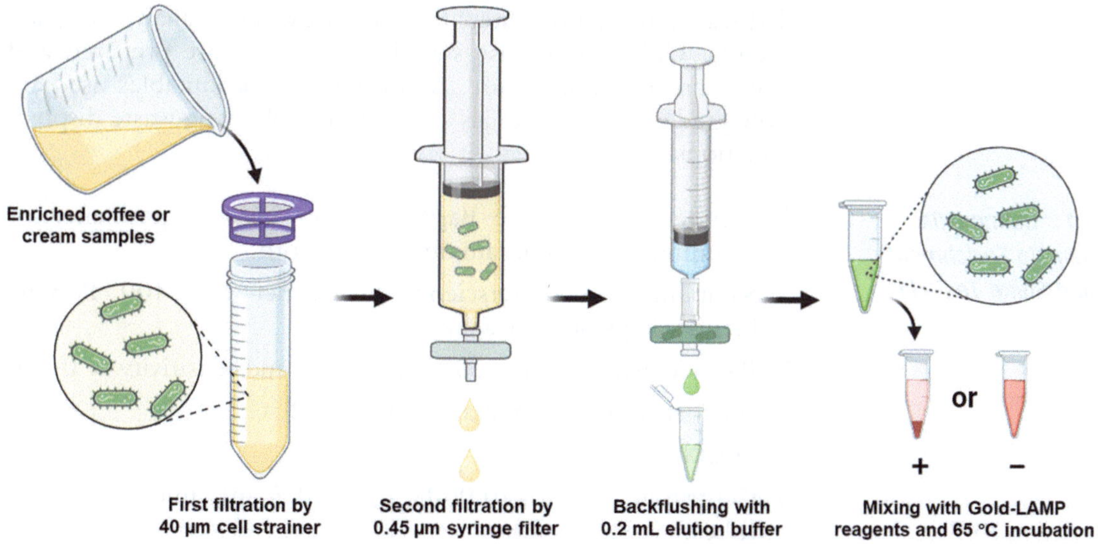

Fig. 2 Schematic illustration of the filtration-based sample preparation steps. First, 10 mL of enriched coffee or cream sample is briefly filtered through a 40 μm cell strainer, followed by a second filtration with 10 mL syringe through a 0.45 μm syringe filter. The syringe filter with remained bacteria is then backflushed with 0.2 mL of elution buffer using a 1 mL syringe through a female-to-female luer connector. A 10 μL of the 0.2 mL bacteria suspension is then mixed with 15 μL of Gold-LAMP reagent mix and incubated at 65 °C. After incubation for 25 min, a negative sample remains as a red dispersion, while a positive sample appears as a red precipitate

Fig. 3 *Salmonella* detection in coffee and cream with filtration-based sample preparation and Gold-LAMP system. NTC, no template control (no *Salmonella* DNA). PC, positive control (10^3 copies of purified *Salmonella* DNA). The photo was taken after LAMP (65 °C for 25 min)

2 Materials

All solutions are prepared using ultrapure DNase/RNase-free distilled water (ultrapure water) from Invitrogen and analytical grade reagents. Unless otherwise stated, all reagents are prepared and stored at room temperature. All disposable consumables used are sterile grade. Waste materials are disposed following waste disposal regulations.

2.1 *Salmonella* Sample Enrichment and Spike Test [11]

1. Blender (Interscience BagMixer 400 SW or equivalent) for blending sample containing lactose broth.
2. Sampling bag (Interscience Stomacher bag 400 Polysilk blender bags or equivalent).
3. Balance, with weights 2000 g capacity and sensitivity of 0.1 g.
4. Incubator, with temperature of 35 ± 2 °C.
5. Heat block, 56 ± 2 °C.
6. Centrifuge (up to 20,000× *g*) for 1.5 mL microcentrifuge tubes.
7. Autoclave for sterilizing broths and utensils.
8. Sterile spoons for transferring food samples. Sterilize by autoclaving at 121 °C for 15 min.
9. Pipettes and pipette tips.
10. Vortex mixer.
11. Coffee and cream samples.
12. Lactose broth (Thermo Scientific Oxoid CM0137B, or equivalent). Dissolve 13 g of lactose broth powder in 1 L of distilled water and mix well. Sterilize by autoclaving at 121 °C for 15 min. This broth can be kept at 4 °C for 2 weeks.

13. *Salmonella enterica* subsp. enterica serovar Typhimurium strain (ATCC 14028 or equivalent) is kept at −80 °C as stock. Grow the strain on tryptone soya agar plate or in tryptone soya broth for 48 ± 4 h at 35 °C incubator.
14. QIAGEN DNeasy Blood and Tissue Kit and absolute ethanol (*see* **Note 1**).

2.2 Filtration of Enriched Coffee and Cream Samples

1. Enriched coffee and cream samples.
2. 40 μm cell strainers from Biosharp.
3. 50 mL centrifugation tubes.
4. 20 mL and 1 mL syringes.
5. 0.45 μm PES syringe filters (13 mm in diameter).
6. Female-to-female coupler luer syringe connectors.
7. Ultrapure DNase/RNase-free distilled water.
8. 1.5 mL microcentrifuge tubes.

2.3 Gold-LAMP Assay

1. Metal cold racks for 1.5 mL centrifuge tubes and 0.2 mL PCR tubes.
2. Ice and ice box.
3. 0.1–10, 2–20, and 20–200 μL pipettes and compatible filtered pipette tips.
4. 1.5 mL microcentrifuge tubes and 0.2 mL PCR tubes.
5. Mini centrifuge for brief centrifugation.
6. Vortex mixer.
7. Marker pens.
8. Weighing balance with sensitivity of 0.001 g.
9. Heat block (65 °C) for 0.2 mL PCR tubes.
10. NanoDrop One Microvolume UV-Vis Spectrophotometer (Thermo Fisher Scientific) for absorbance measurement.
11. PEG/MUA-AuNPs solution (20 nM) prepared as described before [8]. Store at room temperature.
12. 10× isothermal amplification reaction buffer (200 mM Tris-HCl, 100 mM $(NH_4)_2SO_4$, 500 mM KCl, 20 mM $MgSO_4$, 1% Tween 20, pH 8.8 at 25 °C; New England Biolabs B0537S). Store at −20 °C.
13. Betaine solution (5 M; PCR reagent; Merck). Store at 4 °C.
14. Magnesium sulfate ($MgSO_4$) solution (1 M). Dissolve 12.037 g of $MgSO_4$ (Merck M7506) in a final volume of 100 mL of ultrapure water. Sterilize by autoclaving at 121 °C for 15 min. Store at −20 °C.

15. Six specific LAMP primers targeting enterotoxin (*stn*) gene of *Salmonella* spp. [12]:

    FIP: 5′-ACCGGGTGGTAAGCGAATTGCGAGGTTAACCG TCTGGAGC-3′.

    BIP: 5′-TCGGCCTCTTTGGCCATCACTGGCGAAATACT TTGCCGAG-3′.

    F3: 5′-ACCAGATTCAGGGAGTGAGT-3′.

    B3: 5′-CGCGCACGAAATTCGTAAC-3′.

    Loop F: 5′-TGGTAAAGCCCGCGCATCTG-3′.

    Loop B: 5′-GCGCCAGTTCATGCGACTCG-3′.

    Lyophilized and HPLC-purified primers (Integrated DNA Technologies) are reconstituted by adding IDTE buffer (1× TE solution, pH 7.5) to 100 μM according to the manufacturer's instructions. The ready-to-use LAMP primer mixture (220 μL) is prepared by mixing 80 μL of 100 μM FIP/BIP, 10 μL of 100 μM F3/B3, and 20 μL of 100 μM Loop F/Loop B. Store at −20 °C.
16. *Bst* 2.0 DNA polymerase (120,000 units/mL, New England Biolabs M0537M). Store at −20 °C.
17. Deoxyribonucleoside triphosphates (dNTPs) set (25 mM each prepared from 100 mM stock, Thermo Fisher Scientific R0186). Store at −20 °C (*see* **Note 2**).
18. Ultrapure water (Invitrogen). Store at room temperature.
19. Purified *Salmonella* DNA prepared from Subheading 2.1. Store at −20 °C.
20. Mineral oil (BioReagent, for molecular biology, light oil; Merck M5904). Store at room temperature.

3 Methods

3.1 Salmonella *Sample Enrichment and Spike Test [11]*

1. Place the lactose broth at ambient temperature.
2. Add an amount of sample test portion (mass or volume) to a quantity of lactose broth (mass or volume) to yield a tenfold dilution. Place a sterile sampling bag in a container, aseptically weigh 25 g coffee or cream into the sterile sampling bag, and add *Salmonella* ATCC14028 (spiked samples) and 225 mL sterile lactose broth to the bag.
3. Put the bag containing the mixture to blender and blend for 2 min.
4. Incubate the mixture for 24 ± 2 h at 35 °C incubator. The resultant enriched coffee and cream samples are ready for subsequent filtration steps.

3.2 DNA Extraction by QIAGEN DNeasy Blood and Tissue Kit

This section is to obtain purified *Salmonella* DNA as a positive control for subsequent Gold-LAMP assay.

1. Harvest *Salmonella* ATCC 14028 in a 1.5 mL microcentrifuge tube by centrifugation for 10 min at 5000× *g*. Then, discard the supernatant.
2. Resuspend the pellet in 180 μL Buffer ATL (provided by the kit). Add 20 μL proteinase K. Mix thoroughly by vortexing and incubate with a heat block at 56 °C for 30 min.
3. Short spin the mixture, add 200 μL buffer AL, and mix well by vortexing.
4. Short spin the mixture, add 200 μL absolute ethanol, mix well by vortexing, and short spin the mixture.
5. Pipette the mixture into DNeasy mini spin column placed in a 2 mL collection tube (provided by the kit). Centrifuge at 6000× *g* for 1 min. Discard flow-through and the collection tube.
6. Place the DNeasy mini spin column in a new 2 mL collection tube, and add 500 μL buffer AW1. Centrifuge for 1 min at 6000× *g*. Discard flow-through and collection tube.
7. Place the DNeasy mini spin column in a new 2 mL collection tube, and add 500 μL buffer AW2. Centrifugate at 20,000× *g* for 3 min. Discard flow-through and collection tube (*see* **Note 3**).
8. Place the DNeasy mini spin column in a new 1.5 mL centrifuge tube, and pipette 200 μL buffer AE. Centrifuge at 6000× *g* for 1 min to elute DNA from the spin column (*see* **Note 4**).
9. The purified DNA sample is stored at −20 °C (*see* **Note 5**).
10. Measure the absorbances at 260 nm (A_{260}) and 280 nm (A_{280}) of the eluted DNA sample with NanoDrop One Microvolume UV-Vis spectrophotometer.
11. Calculate the mass concentration (ng/μL) based on A_{260} of 1.0 = 50 μg/mL pure double-stranded DNA.
12. Calculate the copy number concentration (copies/μL) based on 1 ng *Salmonella* ATCC 14028 DNA (genome length: 4,964,087) corresponds to 1.52×10^7 copies of the entire genome [13].

3.3 Filtration of Enriched Coffee and Cream Samples

1. Please refer to Fig. 2 for this section (*see* **Note 6**).
2. Place a 40 μm cell strainer on top of a 50 mL centrifuge tube. Pour 10 mL of the enriched coffee or cream sample through the cell strainer. Allow the filtrate to pass through. Discard any residue caught in the strainer.

3. Draw the filtrate collected from the previous step into a 10 mL syringe. Attach the syringe to a 0.45 μm PES syringe filter. Push the filtrate through the filter slowly using the syringe. Keep the syringe filter for the following backflushing (*see* **Note 7**).
4. Connect a female-to-female luer connector to the opposite opening of the syringe filter. Put the syringe filter on top of a 1.5 mL centrifuge tube (the luer connector at the top).
5. Draw 0.2 mL of elution buffer with a 1 mL syringe. Backflush the syringe filter through the luer connector with 0.2 mL elution buffer into the 1.5 mL centrifuge tube. Draw 1 mL air with the 1 mL syringe and flush again (*see* **Note 8**).
6. The backflushed bacteria suspension from coffee or cream is ready for the Gold-LAMP assay. Store at 4 °C for up to several hours (*see* **Note 9**).

3.4 Gold-LAMP Assay

1. Set up a clean work area.
2. Prepare an ice box with ice. Chill metal cold racks on ice.
3. Thaw all frozen reagents (10× isothermal amplification buffer, 25 mM each dNTPs, ready-to-use LAMP primers mixture, and 1 M $MgSO_4$ solution) on a metal cold rack. Mix each thawed reagent well by vortexing. Short spin down the mixture to the bottom of the tube (*see* **Note 10**).
4. Prepare a Gold-LAMP master mix for 4.5 reactions (15 μL for each 25 μL reaction, i.e., 67.5 μL) in a 0.2 mL PCR tube on a cold rack according to the following order: 7.123 μL of ultrapure water, 11.25 μL of 10× isothermal amplification buffer, 4.5 μL of 5 M betaine, 0.675 μL of 1 M $MgSO_4$, 6.3 μL of dNTPs mixture (25 mM each), 4.95 μL of ready-to-use LAMP primers mixture, 32.4 μL of 20 nM PEG/MUA-AuNPs, and 0.302 μL of *Bst* 2.0 polymerase (*see* **Note 11**).
5. Put four PCR tubes on a cold rack. Aliquot 15 μL of the Gold-LAMP master mix into each PCR tube (*see* **Note 12**).
6. Label the four tubes as "NTC," "Coffee," "Cream," and "PC."
7. Add 10 μL of ultrapure water into "NTC" as the no template control. Mix well by vortexing or pipetting. Add 20 μL of mineral oil on top of the mixture.
8. Add 10 μL of the backflushed bacteria suspension from coffee and cream into "Coffee" or "Cream," respectively. Mix well by vortexing or pipetting. Add 20 μL of mineral oil on top of the mixtures.
9. Add 10 μL of purified *Salmonella* DNA (diluted to 10^2 copies/μL from the stock in Subheading 3.1, i.e., 10^3 copies per reaction) into "PC" as the positive control. Mix well by vortexing or pipetting. Add 20 μL of mineral oil on top of the mixture.

10. Incubate the four samples at 65 °C for 25 min (*see* **Note 13**).
11. Take out the samples from the heat block. Interpret the results by the naked eye (Fig. 3) (*see* **Note 14**).

4 Notes

1. Proteinase K provided by the kit is ready to use. For buffer AW1 and AW2 provided, add appropriate volume of absolute ethanol as indicated on the bottle label and shake thoroughly. If there is precipitate in buffer AL, incubate the buffer at 56 °C until it disappears.
2. Avoid repeated freezing/thawing by preparing aliquots.
3. Carefully remove the DNeasy mini spin column from the collection tube. If there is ethanol carryover in the spin column, empty the collection tube, and then reuse it, and centrifuge at 20000× *g* for 1 min to remove the remaining ethanol.
4. The DNA yield can be increased by two times elution, each with 100 μL of buffer AE.
5. Avoid repeated freeze–thaw cycles by preparing aliquots.
6. Make sure proper handling of samples and reagents to prevent contamination. Follow safety protocols while working with bacteria and chemicals.
7. Make sure the filter is attached tightly to the syringe and keep the pushing slow.
8. Make sure thorough rinsing to collect all the bacteria.
9. Mix the suspended bacteria well before adding to the Gold-LAMP reagents.
10. All frozen reagents must be completely thawed and mixed before use.
11. Mix quickly by mild vortexing and spin down the mixture after adding each reagent.
12. Close the PCR tube lid securely to prevent contamination.
13. Put the samples from cold rack to heat block immediately to avoid nonspecific amplification.
14. Do not open the tube of amplified sample, especially in the sample preparation area, to avoid potential carryover contamination.

Acknowledgments

This work was supported by The Hong Kong Polytechnic University, PocNAT Limited, and SGS Hong Kong Limited.

References

1. Kokkinos PA, Ziros PG, Bellou M, Vantarakis A (2014) Loop-mediated isothermal amplification (LAMP) for the detection of *Salmonella* in food. Food Anal Methods 7:512–526
2. Shang Y, Ye Q, Cai S, Wu Q, Pang R, Yang S, Xiang X, Wang C, Zha F, Ding Y, Zhang Y, Wang J, Sun X, Zhang J (2021) Loop-mediated isothermal amplification (LAMP) for rapid detection of *Salmonella* in foods based on new molecular targets. LWT 142: 110999
3. Malorny B, Paccassoni E, Fach P, Bunge C, Martin A, Helmuth R (2004) Diagnostic real-time PCR for detection of *Salmonella* in food. Appl Environ Microbiol 70:7046–7052
4. Zhao Y, Chen F, Li Q, Wang L, Fan C (2015) Isothermal amplification of nucleic acids. Chem Rev 115:12491–12545
5. Kaneko H, Kawana T, Fukushima E, Suzutani T (2007) Tolerance of Loop-mediated isothermal amplification to a culture medium and biological substances. J Biochem Biophys Methods 70:499–501
6. Hara-Kudo Y, Yoshino M, Kojima T, Ikedo M (2005) Loop-mediated isothermal amplification for the rapid detection of *Salmonella*. FEMS Microbiol Lett 253:155–161
7. Azinheiro S, Roumani F, Prado M, Garrido-Maestu A (2022) Rapid same-day detection of *Listeria monocytogenes*, *Salmonella* spp., and *Escherichia coli* O157 by colorimetric LAMP in dairy products. Food Anal Methods 15: 2959–2971
8. Qin A, Fu LT, Wong JKF, Chau LY, Yip SP, Lee TMH (2017) Precipitation of PEG/carboxyl-modified gold nanoparticles with magnesium pyrophosphate: a new platform for real-time monitoring of loop-mediated isothermal amplification. ACS Appl Mater Interfaces 9:10472–10480
9. Schrader C, Schielke A, Ellerbroek L, Johne R (2012) PCR inhibitors–occurrence, properties and removal. J Appl Microbiol 113:1014–1026
10. Cossu A, Witkowsky RD, Levin RE (2014) Fat removal with hydrolyzed corn starch for real-time qPCR detection of *Salmonella enterica* in ground beef in 4.5 hours without enrichment. Food Control 46:475–479
11. Andrews WH, Wang H, Jacobson A, Ge B, Zhang G, Hammack T (2023) BAM Chapter 5: *Salmonella*, US Food & Drug Administration, p 9 https://www.fda.gov/food/laboratory-methods-food/bam-chapter-5-salmonella
12. Srisawat M, Panbangred W (2015) Efficient and specific detection of *Salmonella* in food samples using a *stn*-based Loop-mediated isothermal amplification method. Biomed Res Int 2015:356401
13. Scienceprimer (n.d.). https://scienceprimer.com/copy-number-calculator-for-realtime-pcr

Chapter 20

LAMP-on-a-Chip in Combination with Surface-Enhanced Raman Scattering (SERS) for Pathogen Detection

Verónica Silva, Bernardo A. Nogueira, Sara Abalde-Cela, Juan Luis Paris, and Laura Rodriguez-Lorenzo

Abstract

Loop-mediated isothermal amplification (LAMP) allows the detection of pathogens DNA by designing 4–6 primers that recognize 6–8 distinct regions of target DNA during an amplification reaction. Here we describe a novel microdroplet optofluidic system designed for the sensitive detection of foodborne pathogens, by integrating LAMP with surface-enhanced Raman scattering (SERS). This LAMP-on-a-chip offers the advantage of producing several LAMP reactions inside of the microdroplets using the same quantity of reagents in combination of an indirect SERS detection, utilizing multifunctional gold nanoparticles (AuNPs) and the formation of pyrophosphate generated during DNA amplification in LAMP. The microdroplet-LAMP-SERS methodology detected successfully down to 102.0 pg/μL of target DNA, which was equivalent ~10^2 CFU in 25 mL of UHT milk.

Key words Multifunctional gold nanoparticles, AuNPs, SERS, LAMP, Microdroplets, Glutathione, Microfluidics, Foodborne pathogens, UHT milk, *Listeria monocytogenes*

1 Introduction

Recently, microfluidic platforms, benefiting from miniaturization, automation, and standardization that they offer, have emerged as pivotal instruments for advancing analytical tools for a myriad of research fields and applications [1–3]. Particularly in food safety analysis, these technologies have gained considerable traction, offering portable, cost-effective, and disposable systems that can be integrated for on-site detection measurements with minimal expertise requirements. Among these tools, microdroplets technology [4, 5], a subtype of microfluidics, enables the compartmentalization of chemical and biological reactions within droplets ranging from nanoliter (nL: 10^{-9} L) to picoliter (pL: 10^{-12} L) volumes

The authors Verónica Silva and Bernardo A. Nogueira have equally contributed to this chapter.

Alejandro Garrido-Maestu and Alexandre Lamas (eds.), *Foodborne Pathogens: Loop-mediated Isothermal Amplification Protocols*, Methods and Protocols in Food Science, https://doi.org/10.1007/978-1-0716-4526-0_20,

with fine control over reaction contents [6–10]. This capability makes it a standout choice for their integration with molecular techniques, allowing rapid pathogen detection in food products.

One of these well-known molecular analytical methods is loop-mediated isothermal amplification (LAMP), an isothermal nucleic acid amplification technology. LAMP has been increasingly used as an alternative to PCR, due to its good performance in ultrasensitive detection and biosensing, without the need for harsh temperature cycles that often damage reagents and biological elements under study [11, 12]. LAMP reaction involves DNA polymerase with strong ability to strand displace and a set of four specially designed primers, which recognize a total of six distinct sequences on the target DNA. This reaction can be described as [13]:

$$(\mathrm{DNA})_{n-1} + \mathrm{dNTP} \rightarrow (\mathrm{DNA})_n + \mathrm{P_2O_7}^{4-} \quad (1)$$

During the DNA polymerization by the polymerase (Eq. 1), deoxyribonucleoside triphosphates (dNTP) release pyrophosphate ions ($P_2O_7^{4-}$) as a by-product, which can be indirectly detected [14]. This detection is commonly achieved through turbidity measurements, based on the insoluble complex formed between pyrophosphate and magnesium ions present in the LAMP reaction buffer. Colorimetric methods have been developed using carboxyl acid-functionalized gold nanoparticles (AuNPs), which take advantage of the complexation mechanism involving carboxyl acid molecules on the AuNPs surface, magnesium ions, and pyrophosphate anions generated in LAMP [15, 16]. Although LAMP has been successful integrated into microfluidic devices for detecting foodborne pathogens [17], only few works reported the combination of LAMP with microdroplets technology [18–20]. One of those few publications is the strategy presented in this chapter.

Here, we present an on-chip bioinspired strategy for the detection of *Listeria monocytogenes*, a highly fatal human pathogen [21, 22], by combining LAMP, microdroplet-based microfluidics, and indirect surface-enhanced Raman scattering (SERS) as a transducer [20]. SERS was selected because it is a powerful spectroscopic technique that can be easily integrated with various sampling techniques, especially effective when combined with microfluidics and microdroplets, allowing the development of sensitive, quantitative, and reproducible analytical platforms [23, 24]. Essentially, SERS exploits the amplification of the Raman scattering signal of molecules adsorbed on nanostructured metal surfaces. For that, this indirect SERS strategy is based on the use of multifunctional gold nanoparticles (AuNPs) for DNA detection by reacting with pyrophosphate ions released during DNA amplification in LAMP. The multifunctional AuNPs contain key elements to act as SERS substrate: (1) thiolated poly(ethylene glycol) (TPEG) as stabilizing agent, (2) 1-naphthalenethiol (1-NAT) as Raman reporter, and

(3) glutathione (GSH) as chelating agent of magnesium (II) ions. Thus, the DNA concentration is correlated with the SERS signal of 1-NAT controlled by the aggregation of AuNPs triggered by the complexation of pyrophosphate ions and glutathione with free magnesium ions. This approach enabled the identification of samples containing as little as 102.0 pg/μL of target DNA and the detection of target DNA in food samples artificially inoculated with 3.6×10^2 CFU/mL of *L. monocytogenes* [20]. In this chapter, we describe steps needed to run this LAMP-on-a-microdroplet in combination with SERS using as example the detection of *L. monocytogenes*, which is the pathogen studied in [20].

2 Materials

Analytical grade reagents should be utilized, and all solutions must be prepared using Milli-Q® ultrapure water (18.2 MΩ·cm at 25 °C). Prior to nanoparticle synthesis, ensure thorough cleaning of all laboratory glassware with aqua regia. Strict adherence to waste disposal regulations is essential when handling waste materials.

2.1 Synthesis and Functionalization of AuNPs

Gold nanoparticles (AuNPs) were prepared as described by Turkevich [25]. The reduction of gold in aqueous solution in presence of sodium citrate is produced by the following reaction:

$$
\begin{aligned}
&2HAuCl_{4(aq)} + 3C_6H_8O_{3(aq)}(\text{citric acid}) \\
&\quad \Delta \rightarrow 2Au_{(s)} + 3C_5H_6O_{3(aq)}(3 - \text{ketoglutaric acid}) + 8HCl_{(aq)} \\
&\quad + 3CO_{2(g)}
\end{aligned}
$$

Thus, experimental steps are as follows:

1. Prepare an aqueous solution of sodium citrate (10 mg/mL) and tetrachloroauric acid ($HAuCl_4$) (0.125 M).
2. Add 1000 μL of 0.125 $HAuCl_4$ to 250 mL of ultrapure water, reaching a final gold concentration of 0.5 mM.
3. Stir magnetically and heat the solution up to boiling for 5–10 min. Gold (III) solution is yellow currently.
4. Add quickly 12.5 mL of warmed (50–60 °C) aqueous solution of 10 mg/mL sodium citrate to boiled gold solution under vigorous stirring (*see* **Note 1**).
5. Leave the reaction for 20 min under boiling and vigorous stirring. The color of the solution changes to red, indicating the formation of spherical AuNPs (*see* **Note 2**).
6. Gradually cool down the prepared AuNPs suspension to room temperature (*see* **Note 3**).

For the functionalization to create the multifunctional AuNPs:

1. Prepare 1 mL of 0.25 mM AuNPs in ultrapure water and dilute by 2× the stock dispersion.
2. Add 0.5 μL of 10 mM 1-naphtalenethiol (1NAT) ethanolic solution to 1 mL of the AuNPs suspension and then vortex for 5 s (*see* **Note 4**).
3. Add 50 μL of 1 mM glutathione (GSH) aqueous solution to 1NAT-AuNPs suspension and vortex for 5 s (*see* **Note 5**).
4. Add 21 μL of 1.25 mM thiolated methoxy-poly(ethylene glycol) (Mw 2000 g/mol) (TPEG) aqueous solution to INAT: GSH-AuNPs suspension and vortex for 5 s (*see* **Note 6**).
5. Incubate the resulting INAT:GSH:TPEG-AuNPs at room temperature for 1.5 h under stirring.
6. Centrifuge for 90 min at 4000 rpm.
7. Resuspend the purified samples in 100 μL of ultrapure water.
8. Store final 1NAT:GSH:TPEG-AuNPs at 4 °C until further use.

2.2 Microdroplet Generator and Microdroplet Trapping Reservoir Devices

The microfluidic devices used here are fabricated by photolithographic process, subsequently followed for the fabrication of SU-82025 (photoresist, MicroChem) molds, which are used for polydimethylsiloxane (PDMS) device replication [26]. For the first two steps, a micro- and nanofabrication cleanroom is needed to fabricate SU-8 molds within a class 100 (according to the FED STD 209E, in ISO14644-1 standard in the ISO 5).

The microfluidic device to generate microdroplet as well as the device for microdroplet trapping can be purchased in Dolomite Microfluidics, Fluigent, Darwin Microfluidics, or Sphere Fluidics, among other vendors. The CAD designs of both microfluidic devices can be requested from the authors of this chapter.

In the case of the availability of silicon masters, PDMS replicas can be fabricated as follows:

1. Prepare PDMS (Sylgard 184) by mixing the polymer and the cross-linker at a ratio of 10:1 *wt:wt* (*see* **Note 7**).
2. (Optional if you will not use it immediately) Degas the previously prepared PDMS until the mixture looks homogeneous and no bubbles are being formed (around 40 min in a general-purpose degasser).
3. Pour the degassed PDMS on top of the master and top of an empty petri dish.
4. Degas again the PDMS-coated master until no bubbles are formed (around 40 min in a general-purpose degasser for 3 in. wafer).

5. Cure the PDMS replica contained on top of the master and top of the petri dish for 2 h in an oven at 65 °C.
6. Cut the cured PDMS replica on the master with a scalpel and peel off carefully from the master (*see* **Note 8**).
7. Produce holes for tubing in PDMS replica with a biopsy punch (diameter = 1.25 mm; Kai Medical).
8. Peel off also cured flat PDMS from the empty petri dish.
9. Activate both PDMS pieces, channel containing PDMS device and flat PDMS, by oxygen plasma for 20 s and subsequently bind them for obtaining the final PDMS/PDMS devices (*see* **Note 9**).

2.3 Sample Preparation

2.3.1 Bacterial Culture

We performed and tested the methodology presented here using *L. monocytogenes* WDCM00021 and *L. innocua* WDCM00017 as reference strains [20]. Both strains were supplied by Spanish Type Culture Collection (Parc Cientific de la Universitat de Valencia, Carrer del Catedratic Agustin Escardino Benlloch, 9, 46,980 Paterna, Valencia).

2.3.2 Spiked Samples

After culturing specific reference strain following the standard protocols, buffer solutions or food products (e.g., UHT milk as previously reported [20]) are inoculated with different quantities of bacteria (e.g., 10^1, 10^2, and 10^3 CFU). These samples are used to test DNA extraction kits and then microdroplet-LAMP-SERS methodology presented here. The methodology presented here detects DNA and as a consequence a good DNA extraction must be performed.

This methodology was validated using UHT milk as food matrix. Briefly, 25 mL of UHT milk were mixed with 225 mL of mTa10 broth and subsequently inoculated with 10, 90, and 90×10^2 CFU of *L. monocytogenes.* The mixture was then homogenized for 30 s in a Stomacher 400 Circulator (Seward Limited, UK) and incubated for 25 h at 35 °C. Finally, 1 mL of this mixture was taken for DNA extraction.

The DNA extraction was carried out using the lysis-guanidine isothiocyanate method as previously reported [27, 28]:

1. Centrifuge 1 mL of sample at 13000 rpm for 5 min.
2. Redisperse in 1 mL PBS with Tween 20 and centrifuge once again at 13000 rpm for 5 min.
3. Lyse the bacteria cell pellet using 200 μL of enzymatic solution involving 1 mg/mL lysozyme: 1 mg/mL achromopeptidase in TE 1× buffer.
4. Incubate the mixture at 37 °C under shaking at 1000 rpm for 1 h.

Table 1
Samples analyzed using LAMP-on-a-chip SERS detection [20]

Sample label	Description
Negative control	Water added instead of DNA target in the LAMP reaction
Positive control	10.2 ng/μL target DNA from *L. monocytogenes* in the LAMP reaction
10.2 ng/μL target DNA	10.2 ng/μL target DNA from *L. monocytogenes* in the LAMP reaction
1.02 ng/μL target DNA	1.02 ng/μL target DNA from *L. monocytogenes* in the LAMP reaction
0.102 ng/μL target DNA	0.102 ng/μL target DNA from *L. monocytogenes* in the LAMP reaction
0.0102 ng/μL target DNA	0.0102 ng/μL target DNA from *L. monocytogenes* in the LAMP reaction
L. innocua	13.4 ng/μL DNA from *L. innocua* in the LAMP reaction
SF1—Spiked food	DNA extracted from contaminated 25 mL UHT milk with 10 CFU of *L. monocytogenes*
SF2—Spiked food	DNA extracted from contaminated 25 mL UHT milk with 90 CFU of *L. monocytogenes*
SF3—Spiked food	DNA extracted from contaminated 25 mL UHT milk with 90×10^2 CFU of *L. monocytogenes*

5. Add a strong chaotropic agent (300 μL of 4 M guanidium isothiocyanate and 2% Tween 20) to the mixture.
6. Add 400 μL of the resulting mixture to a tube containing 400 μL of isopropanol to precipitate the DNA.
7. Centrifuge this alcoholic mixture at 13,000 rpm for 10 min.
8. Wash the DNA pellet with 75% isopropanol and centrifuge once again.
9. Dissolve the final pellet in 160 μL of sterile milliQ water by heating at 70 °C for 3 min under shaking (14,000 rpm).
10. Centrifuge the DNA aqueous solution at 13000 rpm for 5 min to remove water-insoluble impurities.
11. Determine the concentration of the DNA extracted using a NanoDrop 2000 °C (Thermo Fisher Scientific Inc., USA).

The summary of the samples analyzed to test the LAMP-on-a-chip SERS detection is shown in Table 1.

2.3.3 Food Samples

Food samples must be prepared according to the ISO and OECD guidelines. The final treatment must allow the DNA extraction to be analyzed with the methodology reported here.

3 Methods

3.1 Preparation of LAMP Reagents

The methodology was tested for the detection of *L. monocytogenes* and therefore, the primers were designed to be selective for this pathogen. The sequences of the primers selected were as follows: hly-F3: TGTGTTTGAGCTAGTGGTTTGG ; hly-B3: CCCATTAGGCGGAAAAGCATAT ; hly-FIP: GCAGCGCTCTCTATACCAGGTACA*tttt*AATG-TCCATGTTATGTCTCCGTTA; hly-BIP: AGGTTTGTTGTGTCAGGTAGAGCG *tttt*CGCTTAAT-AACTGGAATAAGCCAA; and hly-LB: CATCCATTGTTTTGTAGTTACAGAG.

The final volume of the mixture must not be higher than 25 μL (*see* **Note 10**). The reagents must contain the LAMP mixture such as:

1. 15 μL of turbidimetric master mix (OptiGene, Horsham, UK)
2. 1000 nM of FIP/BIP primers (*see* **Note 11**)
3. 200 nM of F3/B3 primers (*see* **Note 11**)
4. 600 nM of LB (*see* **Note 11**)
5. 4 μL of target DNA
6. 4 μL of [Au]=2.5 mM of 1NAT:GSH:TPEG-AuNPs (*see* **Note 12**)

3.2 Microdroplet-Based LAMP Chip

1. Functionalize the channels of the microdroplet generator device by gently flushing Aquapel® manually through a piece of tubing connected to a 1 mL syringe (*see* **Note 13**).
2. Wash the channels with fluorinated oil (HFE 7500 Novec) to remove the remaining Aquapel® and to avoid crystallization and channel clogging.
3. Fill one glass syringe with the dispersed phase, which contain all LAMP reagents mixed as shown in Subheading 3.1 (*see* **Note 14**).
4. Fill another 1 mL glass syringe with the continuous phase, which contain the oil phase, i.e., 2% wt/v Pico-Surf in HFE 7500Novec (Sphere Fluidics, Ltd., Cambridge, UK) (*see* **Note 15**).
5. Assemble each syringe on a syringe pump and connect tubing to inlets of microfluidic generator device.
6. Connect the outlet of microdroplet generator to a microtube (1.5 mL Eppendorf) using tubing.
7. Set syringe pumps: continuous phase (point 4) at 1000 μL/h and dispersed phase (point 3) at 500 μL/h.
8. Start at the same time the syringe pumps to generate the microdroplets that contain the LAMP mixture and wait for

around 5 min for microdroplet stabilization while connecting into waste.

9. Connect the outlet of microdroplet generator to the inlet of the microdroplet trapping reservoir device using the outlet tubing (*see* **Note 16**).
10. Place the full microdroplet trapping reservoir chip in an oven at 62 °C for 60 min to perform LAMP amplification.

3.3 SERS

After DNA amplification by LAMP inside of microdroplets, SERS analysis is performed. An indirect SERS strategy is used following the characteristic peaks, centered at 1553 and 1368 cm^{-1} of the Raman reporter (i.e., 1NAT). The conditions for SERS measurements are as follows:

1. Utilize a Raman spectrometer with a 785 nm excitation laser line. Importantly, a camera should be integrated to observe the microdroplets if a Raman confocal microscope is not available.
2. Collect spectra using a 20× objective with at least a working distance of 2 mm or higher magnification (50× with high working distance) objective (*see* **Note 17**).
3. For each experiment, collect spectra from microdroplets in the reservoirs by acquiring five different spectra from at least five distinct droplets of the same sample (*see* **Note 18**).
4. Preprocess the spectra to correct the baseline and to remove cosmic ray spikes.
5. Calculate the average and standard error of the mean (SEM) of the five spectra from the same sample.
6. Identify the characteristic peaks of 1NAT centered at 1368 and 1553 cm^{-1}, which do not overlap either with master mix Raman peaks or with PDMS peaks.
7. Calculate the area under the curve both 1NAT peaks for all spectra acquire and per sample.
8. Interpret the results:
   - A positive reaction is indicated by the enhancement of SERS signal of 1NAT (i.e., the increase of the peak's intensity) due to the aggregation of 1NAT:GSH:PEG-AuNPs triggered by the complexation of GSH presented on Au surface and pyrophosphate ions that is a by-product of DNA amplification, with free magnesium ions.
   - A negative reaction is indicated by the lack of variation on the SERS signal of 1NAT respect with the negative control (see Table 1).

4 Notes

1. After adding the prepared sodium citrate to the gold solution, sodium citrate turned to citric acid due to pH of gold solution.
2. In the moment of the citrate addition, yellow colored solution suddenly become transparent and colorless. It changes to purple/black and finally to wine red.
3. Store all AuNPs samples at 4 °C until further use. These AuNPs must have an absorbance band centered at 519–522 nm. Transmission electron microscopy analysis should give a primary size in the range of 14–17 nm. The hydrodynamic size estimated by dynamic light scattering should be in the range of 20–30 nm.
4. 1NAT acts as Raman reporter in this method.
5. GSH acts as complexation agent in this method allowing the detection of pathogens using LAMP by aggregation of AuNPs due to the complexation of pyrophosphate and glutathione with free magnesium ions.
6. TPEG acts as stabilizer avoiding the magnesium-conducted aggregation of AuNPs by the formation of the Mg^{+2}-GSH complex as reported previously.
7. The mixing must be vigorous to have a homogeneous mixture at the end. A disposable plastic and fork can be used, as the mixture is sticky after drying. It is possible to prepare a large batch and keep it in the freezer before use.
8. A soft-tip tweezer and isopropanol addition can help in the peeling-off process.
9. Soft pressure by hand should be applied to guarantee the optimal sticking of all the features within the device.
10. If needed, make up 25 μL with sterile, DNAse, RNAse, pyrogen-free ultrapure water.
11. The concentration is in the final volume of 25 μL.
12. The gold concentration must be precisely estimated by UV-Vis spectroscopy [29] or ICP-OES [30].
13. This step is crucial to improve the hydrophobicity of the channels, to encapsulate the LAMP reaction in aqueous droplets surrounded by a carrier oil, as well as to improve the droplet stability during the process. Do not leave the Aquapel® in each microchannel for more than 30 s, as it is hygroscopic and crystalizes in contact with the air.
14. The glass syringe must be of a smaller volume, not more than 1 mL. Ideally, the syringe should contain not more of 25 μL of LAMP reagents (syringes supplied by Hamilton).
15. Plastic syringes can also be used.

16. Stop the generation of the droplets when the reservoir is filled and before the syringe that contains dispersed phase is completely empty (time of injection <3 min). The filling of the reservoir can be followed using an optical microscope.
17. If a portable Raman is used, the working distance is given by the Raman probe (between 4.5 and 12 mm).
18. Conditions for spectra acquisition may vary depending on the Raman spectrometer used (objectives, grating, resolution, total laser power, etc.). However, as a reference, each spectrum can be recorded for 1 s, with five spectra per measurement and a laser power of 70 mW.

Acknowledgments

The authors acknowledge the financial support of the project "Pacto da Bioeconomia Azul," with the reference n.° C644915664-00000026, co-funded by Component C5—Capitalisation and Business Innovation under the Portuguese Resilience and Recovery Plan, through the NextGenerationEU Fund and the project ONE-BLUE (integrated approach to assess the levels and impact of contaminants of emerging concern on BLUE health and biodiversity modulated by climate change drivers), funded from the European Union under the grant agreement no. 101134929. L. R.-L. acknowledges funding from FCT (Fundação para a Ciência e Tecnologia) for the Scientific Employment Stimulus Program (2020.04021.CEECIND). J.L.P. acknowledges grant RYC2021-034536-I funded by MCIN/AEI/10.13039/501100011033 and by European Union Next Generation EU/PRTR.

References

1. Franke TA, Wixforth A (2008) Microfluidics for miniaturized laboratories on a chip. ChemPhysChem 9:2140–2156. https://doi.org/10.1002/CPHC.200800349
2. Ros A (2020) In: Tokeshi M (ed) Applications of microfluidic systems in biology and medicine. Springer. https://doi.org/10.1007/s00216-020-02437-7
3. Watson C, Senyo S (2019) All-in-one automated microfluidics control system. HardwareX 5:e00063. https://doi.org/10.1016/J.OHX.2019.E00063
4. Suea-Ngam A, Howes PD, Srisa-Art M, Demello AJ (2019) Droplet microfluidics: from proof-of-concept to real-world utility? Chem Commun 55:9895–9903. https://doi.org/10.1039/C9CC04750F
5. Ding Y, Howes PD, Demello AJ (2020) Recent advances in droplet microfluidics. Anal Chem 92:132–149. https://doi.org/10.1021/ACS.ANALCHEM.9B05047
6. Huebner A, Sharma S, Srisa-Art M et al (2008) Microdroplets: a sea of applications? Lab Chip 8:1244–1254. https://doi.org/10.1039/B806405A
7. Salmon AR, Esteban R, Taylor RW et al (2016) Monitoring early-stage nanoparticle assembly in microdroplets by optical spectroscopy and SERS. Small 12:1788–1796. https://doi.org/10.1002/SMLL.201503513
8. Abalde-Cela S, Auguié B, Fischlechner M et al (2011) Microdroplet fabrication of silver–agarose nanocomposite beads for SERS optical

accumulation. Soft Matter 7:1321–1325. https://doi.org/10.1039/C0SM00601G

9. Gao R, Cheng Z, Wang X et al (2018) Simultaneous immunoassays of dual prostate cancer markers using a SERS-based microdroplet channel. Biosens Bioelectron 119:126–133. https://doi.org/10.1016/J.BIOS.2018.08.015
10. Choi N, Lee J, Ko J et al (2017) Integrated SERS-based microdroplet platform for the automated immunoassay of F1 antigens in Yersinia pestis. Anal Chem 89:8413–8420. https://doi.org/10.1021/ACS.ANALCHEM.7B01822
11. Notomi T, Okayama H, Masubuchi H et al (2000) Loop-mediated isothermal amplification of DNA. Nucleic Acids Res 28:63e. https://doi.org/10.1093/NAR/28.12.E63
12. Garibyan L, Avashia N (2013) Polymerase Chain Reaction. J Invest Dermatol 133:1–4. https://doi.org/10.1038/JID.2013.1
13. Mori Y, Nagamine K, Tomita N, Notomi T (2001) Detection of loop-mediated isothermal amplification reaction by turbidity derived from magnesium pyrophosphate formation. Biochem Biophys Res Commun 289:150–154. https://doi.org/10.1006/BBRC.2001.5921
14. Fischbach J, Xander NC, Frohme M, Glökler JF (2015) Shining a light on LAMP assays – a comparison of LAMP visualization methods including the novel use of berberine. BioTechniques 58:189–194. https://doi.org/10.2144/000114275
15. Wong JKF, Yip SP, Lee TMH (2014) Ultrasensitive and closed-tube colorimetric loop-mediated isothermal amplification assay using carboxyl-modified gold nanoparticles. Small 10:1495–1499. https://doi.org/10.1002/SMLL.201302348
16. Qin A et al (2017) Precipitation of PEG/carboxyl-modified gold nanoparticles with magnesium pyrophosphate: a new platform for real-time monitoring of loop-mediated isothermal amplification. ACS Appl Mater Interfaces 9(12):10472–10480. https://doi.org/10.1021/acsami.7b00046
17. Trinh TND, Nam NN (2024) Isothermal amplification-based microfluidic devices for detecting foodborne pathogens: a review. Anal Methods 16:1150–1157. https://doi.org/10.1039/D3AY02039H
18. Hsieh SA, Shamsaei D, Eitzmann DR, Anderson JL (2022) Digital droplet Loop-mediated isothermal amplification featuring a molecular Beacon assay, 3D printed droplet generation, and smartphone imaging for sequence-specific DNA detection. Anal Chem 94:11949–11956
19. Hu F, Li J, Zhang Z et al (2020) Smartphone-based droplet digital LAMP device with rapid nucleic acid isolation for highly sensitive point-of-care detection. Anal Chem 92:2258–2265. https://doi.org/10.1021/ACS.ANALCHEM.9B04967
20. Teixeira A, Paris JL, Roumani F et al (2020) Multifunctional gold nanoparticles for the SERS detection of pathogens combined with a LAMP–in–microdroplets approach. Materials 13:1934. https://doi.org/10.3390/MA13081934
21. Allerberger F, Wagner M (2010) Listeriosis: a resurgent foodborne infection. Clin Microbiol Infect 16:16–23. https://doi.org/10.1111/J.1469-0691.2009.03109.X
22. Boelaert F, Tegegne HA, Stoicescu A et al (2019) The European Union one health 2018 Zoonoses report. EFSA J 17. https://doi.org/10.2903/J.EFSA.2019.5926
23. Kant K, Abalde-Cela S (2018) Surface-enhanced Raman scattering spectroscopy and microfluidics: towards ultrasensitive label-free sensing. Biosensors 62(8):62. https://doi.org/10.3390/BIOS8030062
24. Rodríguez-Lorenzo L, Garrido-Maestu A, Bhunia AK et al (2019) Gold nanostars for the detection of foodborne pathogens via surface-enhanced Raman scattering combined with microfluidics. ACS Appl Nano Mater 2:6081–6086. https://doi.org/10.1021/ACSANM.9B01223
25. Enüstün BV, Turkevich J (1963) Coagulation of colloidal gold. J Am Chem Soc 85:3317–3328. https://doi.org/10.1021/JA00904A001/ASSET/JA00904A001.FP.PNG_V03
26. Oliveira K, Teixeira A, Fernandes JM et al (2023) Multiplex SERS phenotyping of single cancer cells in microdroplets. Adv Opt Mater 11. https://doi.org/10.1002/ADOM.202201500
27. Garrido-Maestu A, Azinheiro S, Carvalho J et al (2018) Development and evaluation of Loop-mediated isothermal amplification, and recombinase polymerase amplification methodologies, for the detection of *Listeria monocytogenes* in ready-to-eat food samples. Food Control 86:27–34. https://doi.org/10.1016/J.FOODCONT.2017.11.006
28. Garrido A, Chapela MJ, Román B et al (2013) A new multiplex real-time PCR developed method for *Salmonella* spp. and *Listeria monocytogenes* detection in food and environmental

samples. Food Control 30:76–85. https://doi.org/10.1016/J.FOODCONT.2012.06.029

29. Scarabelli L, Grzelczak M, Liz-Marzán LM (2013) Tuning gold nanorod synthesis through prereduction with salicylic acid. Chem Mater 25:4232–4238. https://doi.org/10.1021/CM402177B

30. Rodriguez-Lorenzo L, Fytianos K, Blank F et al (2014) Fluorescence-encoded gold nanoparticles: library design and modulation of cellular uptake into dendritic cells. Small 10:1341–1350. https://doi.org/10.1002/SMLL.201302889

Correction to: LAMP Detection of Pathogenic *Vibrio* spp. in Seafood and Water Samples Using Colori-Fluorometric Dye

Wataru Yamazaki

Correction to:
Chapter 6 in: Alejandro Garrido-Maestu and Alexandre Lamas (eds.), ***Foodborne Pathogens: Loop-mediated Isothermal Amplification Protocols,***
Methods and Protocols in Food Science, https://doi.org/10.1007/978-1-0716-4526-0_6

The original version of this book was inadvertently published with errors. Typo error in the bottom row of Table 5 in Chapter 6 has been corrected.

Primer mix (for *trh* gene)		
Component	**Amount**	**(μL)**
100 μmol L^{-1} Trh1-FIP	40	μL
100 μmol L^{-1} Trh2-FIP	40	μL
100 μmol L^{-1} Trh1-BIP	40	μL
100 μmol L^{-1} Trh2-BIP	40	μL
100 μmol L^{-1} Trh1-LF	20	μL
100 μmol L^{-1} Trh2-LF	20	μL
100 μmol L^{-1} Trh-LB	20	μL
100 μmol L^{-1} Trh-F3	5	μL
100 μmol L^{-1} Trh-B3	5	μL
Total (for 100 runs)	230	μL

The updated version of this chapter can be found at
https://doi.org/10.1007/978-1-0716-4526-0_6

Alejandro Garrido-Maestu and Alexandre Lamas (eds.), *Foodborne Pathogens: Loop-mediated Isothermal Amplification Protocols*, Methods and Protocols in Food Science, https://doi.org/10.1007/978-1-0716-4526-0_21,

Correction to: LAMP Detection of Pathogenic Vibrio spp. in Seafood and Water Samples Using Colorimetric Dye

[illegible]

Index

Alejandro Garrido-Maestu and Alexandre Lamas (eds.), *Foodborne Pathogens: Loop-mediated Isothermal Amplification Protocols*, Methods and Protocols in Food Science, https://doi.org/10.1007/978-1-0716-4526-0,

MIX
Papier aus verantwortungsvollen Quellen
Paper from responsible sources
FSC® C105338

If you have any concerns about our products,
you can contact us on
ProductSafety@springernature.com

In case Publisher is established outside the EU,
the EU authorized representative is:
Springer Nature Customer Service Center GmbH
Europaplatz 3, 69115 Heidelberg, Germany

Printed by Libri Plureos GmbH
in Hamburg, Germany